Silicon Nanowire Optical Waveguide Sensors: Design and Analysis

Silicon Nanowire Optical Waveguide Sensors: Design and Analysis

DR. RITU RAJ SINGH
Assistant Professor
ECE Department
NSUT, Delhi

CRC Press
Taylor & Francis Group
Boca Raton London New York

CRC Press is an imprint of the
Taylor & Francis Group, an **informa** business

Manakin
PRESS

First published 2026
by CRC Press
4 Park Square, Milton Park, Abingdon, Oxon, OX14 4RN

and by CRC Press
2385 NW Executive Center Drive, Suite 320, Boca Raton FL 33431

CRC Press is an imprint of Informa UK Limited

British Library Cataloguing-in-Publication Data
A catalogue record for this book is available from the British Library

Print edition not for sale in South Asia (Bangladesh, Bhutan, India, Nepal, Pakistan and Sri Lanka).

ISBN: 9781041294788 (hbk)
ISBN: 9781041326052 (pbk)
ISBN: 9781003783589 (ebk)

DOI: 10.1201/9781003783589

Typeset in Times New Roman
by Manakin Press

Manakin
PRESS

Preface

Photonics is transforming how we generate, guide, and use light. Among its many advances, silicon nanowire optical waveguides have emerged as a key building block for compact and highly sensitive devices. Their ability to confine light at the nanoscale makes them especially promising for sensing applications in health, safety, and the environment.

This book, Silicon Nanowire Optical Waveguide Sensors: Design and Analysis, brings together theory, design methods, and applications of these structures. It begins with the fundamentals of silicon photonics and waveguide geometries, then moves to detailed analysis of nanowire waveguides, dispersion behavior, and optimization. The later chapters highlight the applications, including methane detection, glucose sensing, DNA hybridization, and the use of ring resonators for enhanced performance.

The aim is to present clear models supported by numerical validation, while keeping sight of their real-world impact. Readers - whether students, researchers, or engineers - will find both the theoretical foundations and practical insights needed to understand and advance this area of photonics.

It is my hope that this book provides both a reference and a starting point for further exploration into the design of compact, efficient, and versatile integrated optical devices.

Author

Contents

List of Abbreviations

Abbreviation	Meaning	Abbreviation	Meaning
BPM	Beam Propagation Methods	MEMS	Micro-Electro-Mechanical Systems
C-band	Conventional wavelength band	MOVPE	Metal-Organic Vapor Phase Epitaxy
CBE	Chemical Beam Epitaxy	NIR	Near-infrared
CF	Confinement Factor	OAG	Oxide Assisted Growth
CMOS	Complementary Metal Oxide Semiconductor	PBG	Photonic Bandgap
CVD	Chemical Vapor Deposition	PECVD	Plasma Enhanced Chemical Vapor Deposition
D	Dispersion parameter	PIC	Photonic Integrated Circuits
DGMS	Directorate General of Mines Safety	RI	Refractive Index
DWDM	Dense Wavelength Division Multiplexing	RIE	Reactive Ion Etching
EBL	Electron Beam Lithography	RIU	Refractive Index Unit
EIM	Effective Index Method	S-band	Short-wavelength band
EML	Electro-absorption Modulated Laser	SiNWs	Silicon Nanowires
FDM	Finite Difference Method	SNORW	Silicon Nanowire Optical Rectangular Waveguide
FDTD	Finite Difference Time Domain	SOI	Silicon-on-Insulator
FEM	Finite Element Method	SOS	Silicon-on-Sapphire

FSR	Free Spectral Range	STP	Standard Temperature and Pressure
FWHM	Full Width at Half Maximum	TE	Transverse Electric
GVD	Group Velocity Dispersion	TM	Transverse Magnetic
IC	Integrated Circuit	TMM	Transfer Matrix Method
L-band	Long-wavelength band	VLS	Vapour Liquid Solid
LED	Light Emitting Diode	VLSI	Very Large Scale Integration
LPE	Liquid-Phase epitaxy	WDM	Wavelength-Division Multiplexing
MBE	Molecular Beam Epitaxy	ZDWs	Zero-Dispersion Wavelengths
Au	Gold	F	Finesse
DNA	Deoxyribonucleic Acid	n	Refractive index
dsDNA	double strand DNA	Q	Quality-factor
EDTA	Ethylene-diamine-tetraacetic Acid	Sh	Waveguide sensitivity
H_2O	Water	v_g	Group velocity
HF	Hydrofluoric	v_p	Phase velocity
NaF	Sodium fluoride	β	Propagation constants
Si	Silicon	Γ_s	Slot confinement factor
SiO_2	Fused silica	Γ_w	Waveguide confinement factor
ssDNA	single strand DNA	λ	Wavelength

1 Introduction to Silicon on Insulator Waveguides

1.1 SILICON PHOTONICS

Silicon Photonics that became reality in today's technology had its foundations in early 80's. During this period, silicon photonics started to gain popularity and became a hot topic of research across the globe. It's huge prospective was predicted first in 1985 with the study of waveguides in Silicon-on-insulator (SOI) technology (Soref and Lorenzo, 1985; Reed, Headley and Png, 2005). It was commercially implemented in 1989 by Bookham Technology Ltd (Rickman, 2014). The next decade witnessed the emergence of silicon based sensor technology having wide applications in the field of integrated gyroscopes and pressure sensors (Mosser *et al.*, 1991; Maenaka *et al.*, 1996). Products based on wavelength division multiplexing were also reported in the late 1990s (Thomson *et al.*, 2016). A silicon photonics based company, Luxtera, became pioneer in developing photonic devices with CMOS process in 2001 and a silicon photonic system with only passive devices was fabricated in 2002 (Lawrence C. Gunn, 2002). This work was further carried on by Intel that fabricated chips inlaid with phase modulators and splitters. A single 40 Gbps chip with both transmitter and receiver operating at 40 Gbps were fabricated in 2007 by Luxtera (Hochberg and Baehr-Jones, 2010).

Most of the advancements in this field happened in the past decade itself. It is now used for more cost and power efficient data communication systems. Optical modulators were further explored, albeit studied since mid-1980s (R. A. Soref and Bennett, 1987; R. Soref and Bennett, 1987), their speed of execution was quite low to be practically used. However, in recent studies plasma dispersion effect promised better modulator performance speed wise (Thomson *et al.*, 2016). Photonic devices like photodetectors, splitters, interferometers, spectrometers, and others, started to come into picture. New materials started to emerge as alternatives like silicon-on-sapphire (SOS) (Boeuf *et al.*, 2015), silicon-on-silicon nitride, silicon-on-porous silicon, suspended silicon (Soler Penadés *et al.*, 2014) and germanium-on-silicon (Nedeljkovic *et al.*, 2015). Hereafter, mass focus started to shift to non-linear effects in SOI. With the new direction more capabilities of silicon have been exploited. It can now be seen in lasing applications, wavelength conversion and amplification too (Jalali, Paniccia and Reed, 2006). To make silicon photonic a reality in day-

DOI: 10.1201/9781003783589-1

to-day life and commercially viable, it needs to be compatible with existing VLSI technology. Along with the basic optical applications, it has scope in mid-IR range of devices. This possibility came into light due to the thermal properties of silicon and its ability to sustain optical damage up to an appreciable verge (Weakliem and Redfield, 1979; Silva *et al.*, 2001).

Silicon photonics evolved as a result of various advantages of silicon such as: easy availability as superior quality crystal, easy compatibility of tiny chip, and low cost material which is foreseen to be low cost silicon photonics. The silicon-on-wafer has the added quality of providing high optical confinement between Si and SiO_2, which makes nanometer level photonic devices possible. Si has a refractive index of 3.45 and SiO_2 has 1.44. This great gap in terms of optics makes the nonlinear applications like Raman and Kerr effects possible too. Silicon has even more advantages in terms of thermal conductivity and damage threshold for optical practices which are both about ten times more than that of GaAs. Optical nonlinearities like Kerr effect etc., are magnified by 100 times and Raman Effect by 1000 times that of silicon based fiber. It has a wavelength range of 1.1μm to 7μm for low loss effect which is more than the conventional communication band of 1.3μm to 1.55μm. It implies Silicon photonics is not limited to near-infrared (NIR) region (Jalali *et al.*, 2006; Jalali, Paniccia and Reed, 2006) only.

1.2 PHOTONIC INTEGRATED CIRCUIT

Everything in the photonics world from the detection of light in semiconductors in 1883 (Loebner, 1976; Holonyak, 1997) to the invention of transistors in 1947 (Bardeen, 1988; Ross, 1998; Riordan, Hoddeson and Herring, 1999), it's been a rather fascinating progressive journey to the development of photonic integrated circuits. The semiconductor injection lasers invented in 1962 was a product of three years' worth of research from 1959 on electronic integrated circuits (ICs) followed by invention of Lasers in 1960. The III-V alloy injection lasers were proposed by Holonyak and Bevaqua in 1962 (Holonyak and Bevacqua, 1962). When heterojunctions were added to the concoction, it made optical and carrier confinement possible. In 1969, a double heterojunction injection laser was advanced which worked at room temperature, first of its kind, and it also provided continuous wave operations for semiconductor lasers. The theory behind photonic ICs was a result of the previous inventions and the fact that in the presence of an externally applied voltage, III-V semiconductors can modulate light. Its application was defined by Kao and Hockham with their contribution of fibers with low loss in optical frequency range (Kao and Hockham, 1966).

By the end of 1960s, concepts for foundation of photonic integrated circuits were boned up (Kish *et al.*, 2018). However, the commercial application was still not possible because of several factors like diverse requirement of devices for its implementation, issues with scaling, etc. as described by Kaminow (Kaminow, 2008). So LEDs and Lasers weren't seen in telecommunication networks till 1980 and 1984 respectively. For the next decade, the focus was on optimizing these systems in terms of data rates. It grew with about twice per 2.2-year rate. Mostly the issue revolved around the cost of such systems. The integration was certainly beneficial in terms of various significant factors but the cost optimized system was first witnessed in 1986 with two applications on one chip. The electro-absorption modulated laser (EML) hence realized was successful in terms of better bandwidth, reduction of transmission impairments and lesser cost than electrical counterparts (Kish *et al.*, 2018). Towards the end of the millennia, wavelength division multiplexing employing systems were worked on. Tunable LASERs along with optical add/drop multiplexers provided the WDM operations, which resulted in the major progression, widely in tunable lasers (Coldren *et al.*, 2004).

1.3 PHOTONIC WAVEGUIDE

An optical waveguide is a physical structure that guides electromagnetic waves of the optical spectrum. Common types of optical waveguide include optical fibers and integrated waveguides. Optical fibers are mainly used as transmission medium for long haul optical communication. However, integrated waveguides are used as guiding component in integrated optical circuits for ultrashort optical transmission. For an optical waveguide designer, waveguides play the same role as paintbrushes do for a painter, always use the one that better fits your project, and the first choice to be done is adjustment of the refractive index contrast of the waveguide (Melloni *et al.*, 2009). The optimization of refractive index contrast improves the characteristics of light propagation and the scale of integration of photonic devices. This choice is often tricky, because every optical technology is a world apart, with its own rules, advantages and limitations.

1.3.1 Rectangular Waveguide

The rectangular dielectric waveguide is one of the most commonly used structure in integrated optics. Demand of high speed backplanes in integrated electronic and photonic integrated chips served as a driving force in search of materials and robust fabrication techniques for planar waveguide structures (Marcatili, 1969). The slab and circular waveguides are not practical to use on a substrate as: the slab waveguide has no lateral confinement (Kogelnik, 1975), and the circular Fibre is not compatible with the planar processing technology being used to make planar structures. However, the rectangular waveguide is the natural structure with two dimensional confinement shown in Fig. 1.1.

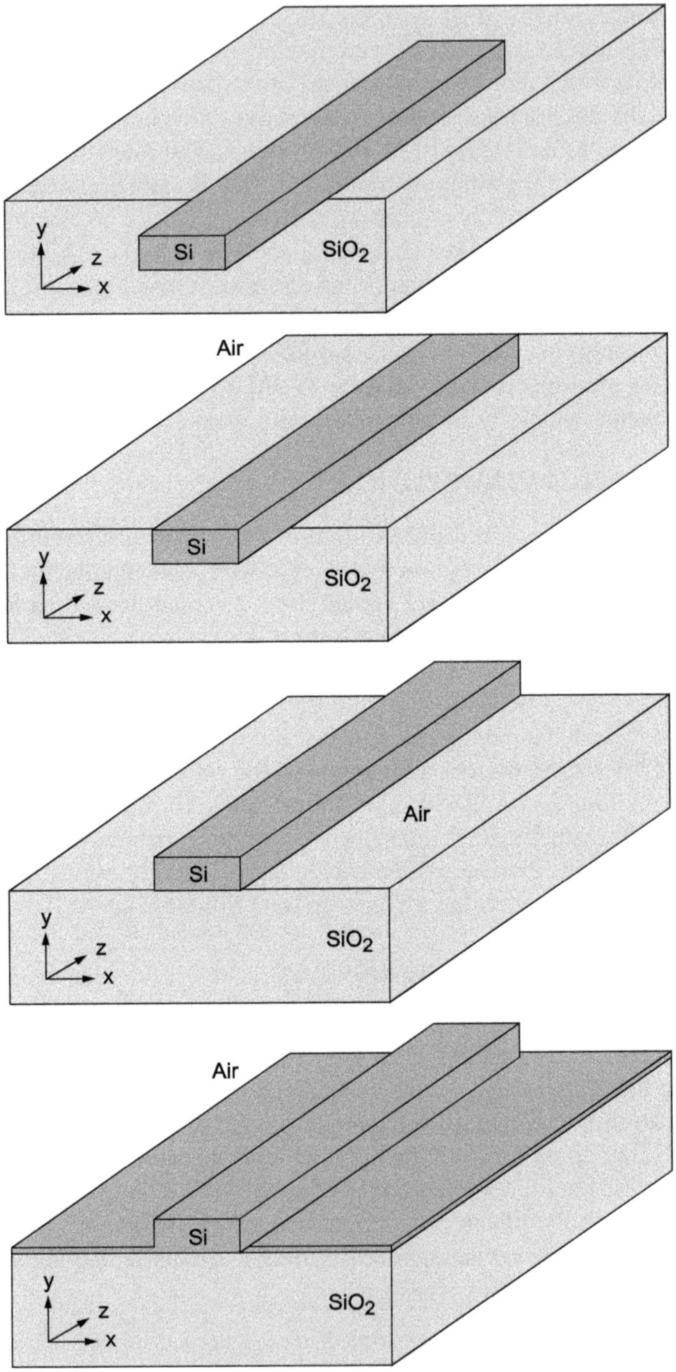

Fig. 1.1. Configurations for rectangular waveguides (a) buried, (b) diffused, (c) ridge, and (d) rib.

Fig. 1.1 shows four types of rectangular waveguide that can be employed in an integrated optical circuit. The four types of waveguides illustrated are: the buried waveguide, diffused waveguide, surface ridge waveguide, and the surface rib waveguide. These geometries are relatively simple to fabricate than other waveguide structures using standard lithographic and overgrowth techniques. By cognizance of the fact that, the index of the rectangular region must be larger than the surrounding medium for the structure to guide light.

One of the basic light guiding elements of silicon photonics is optical waveguide structures; ridge and rib waveguides (Pollock and Lipson, 2003b). These are popular and extensively studied over the years. At present, they are being used for different applications i.e; lasers, amplifiers, modulators and switches. Rectangular waveguide are the essential parts to deliver photons and provide host medium to light matter interaction. The ridge waveguides generally consist of two parts; a silica substrate and a rectangular silicon ridge part, the rib waveguides consist of three parts; a silica substrate, a thin silicon layer on top of the substrate and a rectangular silicon rib part. The index of the rectangular region must be larger than the surrounding medium for the ridge and rib waveguides in order to guide light based on index contrast confinement mechanism. In addition, the waveguides must be designed to operate in fundamental mode, otherwise the higher-order modes will couple into the neighbour regions, that causes a high propagation loss. Besides, higher order modes are prone to dispersion that dilates optical pulses while propagating down to waveguide due to mode dispersion. Hence, the generalized conditions for fundamental modes are important to discuss. Ridge waveguide is a superposition of a strip section and a planar layer. The strip part is the core part with a high index of refraction and planar part is a substrate with low refraction index placed under the strip (Eti and Kurt, 2016). Rib waveguides, which has been studied as a structure similar to ridge waveguide. However, the bottom thin layer of core part is expanded partially to planar layer with the same refractive index.

1.3.2 Slot Waveguide

Silicon slot waveguides are formed from a low index medium surrounded by a high index medium. Light is confined in the low index platform due to the electric field discontinuity at the interface between high-index-contrast materials. Such field guidance is not common in the standard optical waveguides. In recent years several attempts were made to modify the basic structure to confine light at low index medium. The field handling capabilities of the slot optical waveguide have initiated huge research activities in the field of sensors, modulators, couplers, lasers and filters. At all these applications, one can get the benefit of strong field storing in the low index media. That is the main advantage of slotted-waveguides compared to conventional dielectric waveguides (Anderson, Schmidt and Lipson, 2006). The behaviour of guiding optical intensity inside slot region totally depends upon the slot material, this property lead to better sensing applications as compared to conventional rectangular waveguide. Even though vertical voids are common in use by adding vertical, horizontal or cross type air slots, one can change the field confinement mechanism and polarization supported by the waveguide. It is found that adding slots can modify the nature of the mode supported in the waveguides. Thus slot waveguide will be a better option to design various photonic devices.

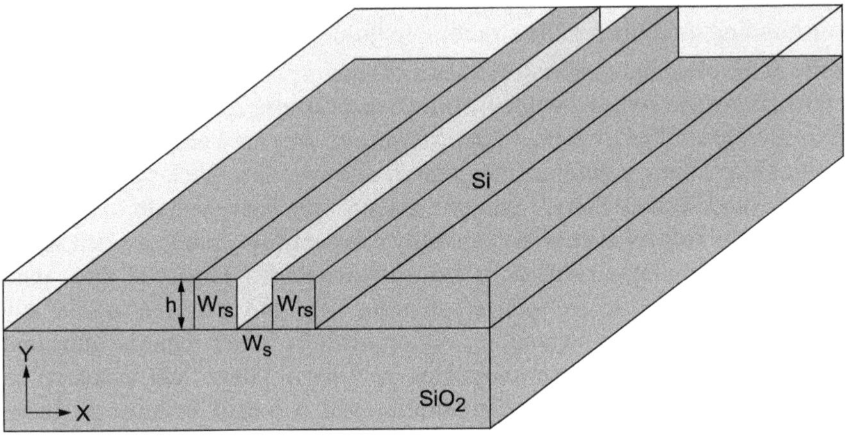

Fig. 1.2. Schematic of a straight slot waveguide.

Generally, the low index media is selected as air, which makes it possible to carry high intensity light sources in air or to alleviate sensitive light interaction with the deposited material for sensing applications. According to the application areas of the slot waveguides, there could be certain advantages compared to rectangular waveguide, such as increasing the area of the low index medium over the total cross sectional area of the waveguide.

1.3.3 Silicon Nanowire Optical Rectangular Waveguide (SNORW)

A fused silica (SiO_2) substrate of waveguide has refractive index of 1.444 at 1550 nm. Over the substrate, silicon wires are vertically arranged in periodic array pattern. These silicon wires are of few nanometers in diameter, which is having period of subwavelength range. The height of these nanowires depend upon the silicon fabrication technology, accordingly, 220 nm of height is taken in further analysis as per current fabrication approaches. The refractive index of silicon is 3.4752 at 1550 nm wavelength. The arrangement of silicon wires over fused silica as a substrate, will make the whole structure compatible to SOI technology. The total width of waveguide '*w*' and height '*h*' of wires is shown in Fig. 1.3. The difference in refractive index along the lateral direction, between the silicon wires and fused silica, allows the guidance of transverse optical signal. Since, it is possible to guide wave in this structure, it is named as Silicon Nanowire Optical Rectangular Waveguide (SNORW). The goal of SNORW structure is to increase the optical confinement within the silicon nanowires. With SiO_2 as a cladding material, SNORW can be used for communication purpose. On the other hand, with different variety of cladding, SNORW can be used for sensing applications. If silicon nanowires inside SNORW are placed close to each other, the average refractive index will become higher than surrounding, with optical mode can be seen in lateral direction, thereby guiding light. This SNORW structure may look similar to the photonic bandgap (PBG) structures such as photonic crystal waveguide. However, in SNORW, there is no bandgap as in photonic bandgap structures. The individual nanowires have scattering points, which are much lesser than that of PBG, at least 5 to 10 times smaller. For the SNORW structure to work efficiently with low radiation loss, an individual nanowire should not refract light. This can be achieved if the alteration in the face front of an optical wave does not occur while propagating through silicon nanowire forest.

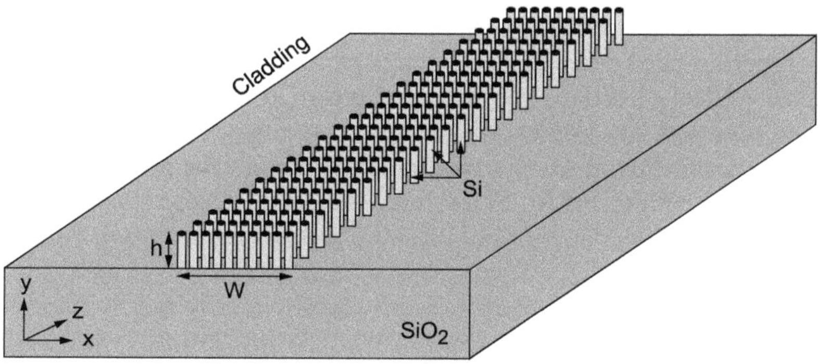

Fig. 1.3. Schematic diagram of SNORW.

1.4 SILICON NANOWIRE FABRICATION APPROACHES

Silicon nanowire optical rectangular waveguide consist of periodic silicon nanowires in an arrayed pattern placed in a close proximity. Such structures can guide an optical mode provided that the electric field is polarized along the length of the nanowires (Khorasaninejad and Saini, 2010). The two approaches that are available to fabricate nanowire waveguides are top-down and bottom-up approaches. In top-down approach, masking is used to etch out the undesired void region resulting to vertically grown silicon wires. The shape of wire depends upon the masking properties. However, bottom-up approach is preferred to fabricate devices of nanoscale dimensions. In bottom up approach, silicon nanowire waveguide is fabricated by growing an array of silicon nanowires on silicon oxide substrate and then undesired portion of the silicon wires are etched (Khorasaninejad and Saini, 2010). There are several methods to create semiconductor nanowires including Silicon Nanowires (SiNWs). These methods can be categorized into two main approaches: (1) Bottom-up approach (growth process) and (2) Top-down approach (etching process).

Table 1.1. Fabrication approached for silicon nanowire.

Bottom-Up Approach	Vapor Liquid Solid Growth
	Oxide Assisted Growth
	Template-Directed Synthesis
Top-Down Approach	Metal-nanoparticle Catalyzed Chemical Etching
	Lithography followed by Reactive Ion Etching

1.5 BOTTOM-UP APPROACH

1.5.1 Vapour Liquid Solid Growth

This is a most common method initially discovered in 1964 by Wagner and Ellis. Vapour Liquid Solid (VLS) growth is a general method for the synthesis of SiNWs, using this, accurate physical dimensions and high quality electronic and optical properties can be obtained. Silicon is an isotropic crystal whose growth is required in vertical direction. The wires are grown vertically upward by creating predefined sites with Au droplets which also act as catalytic material. Silate gas is sputtered to create alloy of Au-Si nanodroplets by raising the reaction temperature over eutectic temperature of Au-Si mixture. The detailed steps (Wagner and Ellis, 1964; Wagner and Ooherty, 1968) of this VLS process are shown in Fig. 1.4.

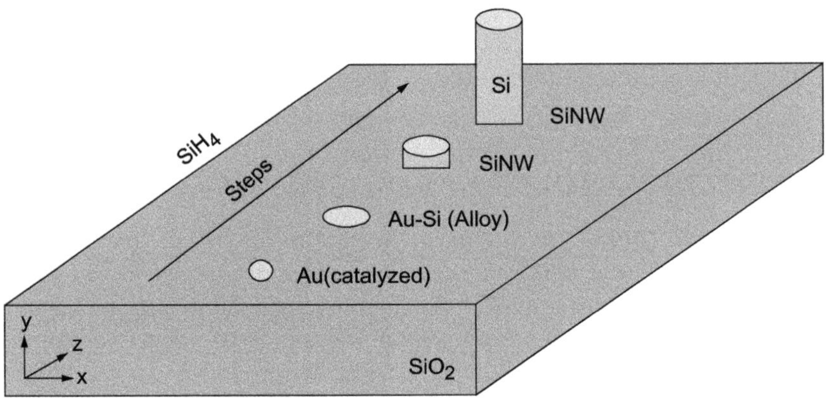

Fig. 1.4. Steps of VLS mechanism.

1.5.2 Oxide Assisted Growth

In previous technique, Au nanodroplets act as a metallic material due to which some metal impurities exist in the grown nanowires (Schmidt *et al.*, 2009). To remove this fault, a new method was proposed by various researchers which is known Oxide Assisted Growth (OAG) technique. By applying this method, silicon nanowires of diameter ranging from 6 to 28 nm can be achieved (Zhang *et al.*,

2000), which is very difficult to grow in VLS method. However, this method is quite similar to VLS techniques, the only difference is the use of oxide which plays the same role of growth instead of Au nanodroplets. This oxide act as key sites for starting point of nanowire growth. Key part of the process is generation of silicon monoxide vapour that can be done either by thermal evaporation (Wang *et al.*, 1999) or laser ablation (Wang *et al.*, 1998). The silicon formation occurs in two essential steps as follow: $Si_xO \rightarrow Si_{x-1} + SiO$ ($x > 1$) and $2SiO \rightarrow Si + SiO_2$. These decompositions happen at temperature of 930°C creating nucleation of silicon nanoparticles. These nuclei of silicon are cladded by shells of silicon oxide and act as seeds for growth of SiNWs.

1.5.3 Template-Directed Synthesis

Multilayer nanowires are also a challenging tasks, which is very difficult to achieve in VLS and OAG methods. However, multi-layered nanowires (Routkevitch *et al.*, 1996) can be achieved by the technique named as Template-Directed Synthesis. In this method, size and shape of nanowires can be controlled with the size and shape of template. Uniformity and diameter of each nanowire can be controlled with this template-directed synthesis. Direct deposition of material onto template yield the formation of nanowires. In addition, single crystal semiconductor nanowires can also be formed by combining this Template-Directed method with VLS or OAG methods.

1.6 TOP-DOWN APPROACH

1.6.1 Metal-nanoparticle Catalysed Chemical Etching

Instead of growing rods in cover region, this method is used to create holes in substrate. In other words, this is known as porous nanowires having nano-holes on substrate material. In this method, metal particles are deposited first on the surface of silicon bulk material to enhance its dissolution. The layer of metal particle can be formed with methods such as electrochemical deposition and thermal evaporation sputtering. The created sample is then immersed inside Hydrofluoric (HF) solution for electrochemical dissolution. In HF solution, etching process occurs at the position where metal particles are present. The metal particles shrink into substrate, and the resultant nano-holes are formed inside silicon bulk material. The depth of hole depends on the etching duration and the diameter of the hole depends on the diameter of metal nanoparticles.

1.6.2 Lithography followed by Reactive Ion Etching

The most controlled way of top-down process is using lithography to define a mask pattern followed by Reactive Ion Etching (RIE). In semiconductor processing lithography is used to transfer a design pattern from a mask to a wafer. In typical optical lithography the pattern size is in a few micrometers that is sufficient for most of the applications. However, in nanowire structures, sizes are extended from hundreds of nanometer to tens of nanometer pushing us to look for higher resolution methods of lithography. Electron Beam Lithography (EBL) is basic pattern transfer tool used in nanotechnology area. After defining the pattern by EBL, the next step is etching. Generally, etching occurs in two directions simultaneously, vertical and lateral. To have a high aspect ratio the vertical etching should dominate. Especially, for nanowire formation since the diameter is in range of nanometer while the length can be few of micrometers there is a need to decrease lateral etching as much as possible. This goal can be obtained by utilizing a highly anisotropic etching in which the lateral etch component is very small resulting in a faithful pattern transfer. On the other hand, diameter and length of nanowires are mostly independent by using anisotropic etching.

❑❑❑

2 Theoretical Analysis of Rectangular Waveguide

Photonic integrated circuits have received noticeable attention in the last two decades due to its capability to integrate various optical components on a single substrate. The PICs are analogous to electronic integrated circuits except that photons are controlled instead of electrons. The feasibility of PIC is possible because of the availability of extremely advanced substrate processing techniques, e.g., lithography, CVD, LPE, etc. Now-a-days, silicon photonics is widely accepted key technology for next generation communication system. Since, the processing of silicon has been developed few decades ago by the electronics industry to a level that is more than sufficient to be well understood by the people (Reed and Knights, 2004). Two main reasons for choosing silicon as an integrated optical material are technological and economical. Some other advantages include; optical transparency at 1330 nm and 1550 nm wavelengths, electronic and optical integration, availability of stable native oxide for isolation, etc (Reed and Knights, 2004; Soref, 2006). Silicon waveguides provide a platform for highly integrated optical systems. The high refractive index contrast in these systems allow high degree confinement with extremely high possibility of compact structures. In addition, cover material also plays a vital role for guiding phenomenon as the intensity of evanescent field can be altered by varying the refractive index of waveguide cover material. The intensity varying in the cover region will play a prominent role for low index guiding structures i.e., slot waveguide, photonic crystal waveguide, etc. Since the modal behaviour of guiding structure depends upon the waveguide structural parameters and refractive index distribution along the waveguide cross-section, the investigation on modal properties of different waveguide geometry is necessary before implementing it to any application, which is further discussed in this chapter.

This chapter is organized to give a brief review of theoretical modal investigation of generalized field equations starting from basic Maxwell's equations transforming to generalized TE and TM modes for guiding phenomenon. The mode equations are derived for buried rectangular waveguide, explaining the dependency of higher order modes on waveguide parameters with the help of both theoretical and analytical methods. The mode profile of rectangular waveguide is further shown graphically, where guiding inside higher refractive indexed material is achieved.

DOI: 10.1201/9781003783589-2

2.1 MAXWELL'S EQUATIONS FOR MODE CALCULATION

Optical waveguides are key elements of photonic devices that perform guiding, coupling, switching, splitting, multiplexing and de-multiplexing of optical signals. Passive waveguide components can be integrated into a single chip using planar technology which is similar to microelectronics. Although the operation of waveguide device is well researched and understood, their performance relies on various parameters i.e. geometry, wavelength, material profile, and field distribution. In this section, the goal is to determine the generalized mode expression which can be applied to different optical SOI waveguides. To begin the analysis, a wave equation is developed. In differential form, well known Maxwell's equations can be written as:

$$\nabla \times \vec{E} = -j\omega\mu_0\vec{H} \qquad ...(2.1)$$

$$\nabla \times \vec{H} = j\omega\varepsilon\vec{E} + \vec{J} \qquad ...(2.2)$$

$$\nabla \cdot \vec{D} = \rho \qquad ...(2.3)$$

$$\nabla \cdot \vec{B} = 0 \qquad ...(2.4)$$

Where, E is the electric field intensity (V/m), H is magnetic field intensity (A/m), D is electric flux density (C/m^2), B is magnetic flux density (A/m^2), ρ is electric charge density (C/m^3), μ_0 is permeability of free space (H/m), J is current density (A/m^2), ε_0 is permittivity of free space (F/m), ε_r is relative permittivity of the medium ($\varepsilon_r = n^2$) and ω is angular frequency (rad/sec). Here, the propagation of light in the waveguide is investigated without worrying about the origin of the light inside the waveguide. So, the Maxwell's equations which govern the electromagnetic waves inside the waveguide are assumed as source free. Thus, the charge density (J) and the conduction current densities (ρ) in the present case are zero. Equation 2.2 can also be written in form of refractive index (n) as

$$\nabla \times \vec{H} = j\omega\varepsilon_0 n^2\vec{E} \qquad ...(2.5)$$

The differential operators become $d/dy = 0$, $d/dz = -j\beta$, and $d/dt = j\omega$. By substituting these differential operators in Eq. 2.1 and equating the individual x, y, z components, the following equations are generated as:

$$j\beta E_y = -j\omega\mu_0 H_x \qquad ...(2.6)$$

$$j\beta E_x + \frac{\partial E_z}{\partial x} = j\omega\mu_0 H_y \qquad ...(2.7)$$

$$\frac{\partial E_y}{\partial x} = -j\omega\mu_0 H_z \qquad ...(2.8)$$

Similarly, by substituting differential operators in Eq. 2.2 and equating the individual components will result to the following equations

$$j\beta H_y = j\omega\varepsilon_0\varepsilon_r E_x \qquad ...(2.9)$$

$$j\beta H_x + \frac{\partial H_z}{\partial x} = -j\omega\varepsilon_0\varepsilon_r E_y \qquad ...(2.10)$$

$$\frac{\partial H_y}{\partial x} = j\omega\varepsilon_0\varepsilon_r E_z \qquad ...(2.11)$$

2.2 TRANSVERSE MAGNETIC (TM) MODE

In transverse magnetic mode, H_x, E_y, $H_z = 0$ and E_x, H_y, $E_z \neq 0$. Also, H_y is continuous in TM mode, therefore, by substituting the value of E_x and E_z from Eq. 2.9 and Eq. 2.11 to Eq. 2.7, relation with only H_y component is obtained as:

$$\frac{\partial^2 H_y}{\partial x^2} + (\omega^2\mu_0\varepsilon_0\varepsilon_r - \beta^2)H_y = 0 \qquad ...(2.12)$$

where, β is the propagation constant. It is already known that, $\mu_0\varepsilon_0 = 1/c^2$; $\varepsilon_r = n^2$ and $\omega^2/c^2 = k_0^2$, where k_0 is the free space wave number. On substituting these expressions to Eq. 2.12, TM mode equation will be achieved as:

$$\frac{\partial^2 H_y}{\partial x^2} + (k_0^2 n^2 - \beta^2)H_y = 0 \qquad ...(2.13)$$

The TM modes are the transverse magnetic modes where the component of the magnetic field vanishes along the waveguide ($H_z = 0$). At boundary parallel component of electric field (E_z) is always continuous. Thus, the boundary condition, which is H_y and E_z should be continuous at the boundary of waveguide.

2.3 TRANSVERSE ELECTRIC (TE) MODE

In transverse electric mode, H_x, E_y, $H_z \neq 0$ and E_x, H_y, $E_z = 0$. Since, E_y is continuous in TE mode, therefore by substituting the value of H_x and H_z from Eq. 2.6 and Eq. 2.8 to Eq. 2.10, relation for TE mode is obtained as:

$$\frac{\partial^2 E_y}{\partial x^2} + (k_0^2 n^2 - \beta^2)E_y = 0 \qquad ...(2.14)$$

The TE modes are the transverse electric modes, where the component of the electric field vanishes along the waveguide ($E_z = 0$). At boundary parallel component of magnetic field (H_z) is always continuous. Thus, the boundary condition, which is E_y and H_z should be continuous at the boundary of waveguide.

2.4 ORTHOGONAL STRUCTURE FOR RECTANGULAR WAVEGUIDE

In this section, parametric analysis has been investigated theoretically for a SOI based rectangular optical waveguide. This section is mainly concentrated on the modal behavior of electric and magnetic field pattern in a rectangular optical waveguide. Here, SOI based structure is used to design rectangular optical waveguide. In which silicon oxide (SiO_2) and silicon (Si) with different refractive index in vertical direction allows the optical signal to propagate in transverse direction. A rectangular optical waveguide of width ($w_r = 2a$) and height (h) of 220 nm has been considered as shown in Fig. 2.1(a). The refractive index of waveguide (n_h) and cladding (n_c) are 3.4752 and 1.444, respectively at operating wavelength (λ_0) of 1550 nm.

Analysis of optical waveguides are possible by various analytical methods such as Marcatilli's Method (Westerveld et al., 2012), Modified Effective Index Method (EIM) (Wang, Farrell and Freir, 2006), Semi-analytical approximations (Gehlot and Sharma, 2014) and numerical methods like Finite Difference Method (FDM) (Lusse *et al.*, 1994), Finite Element Method (FEM) (Rahman and Davies, 1984), Beam Propagation Methods (BPM) (Chung and Dagli, 1990, Finite Difference Time Domain (FDTD) Method (Yang et al., 2001), and Transfer Matrix Method (TMM) (Hong, Huang and Makino, 1992).

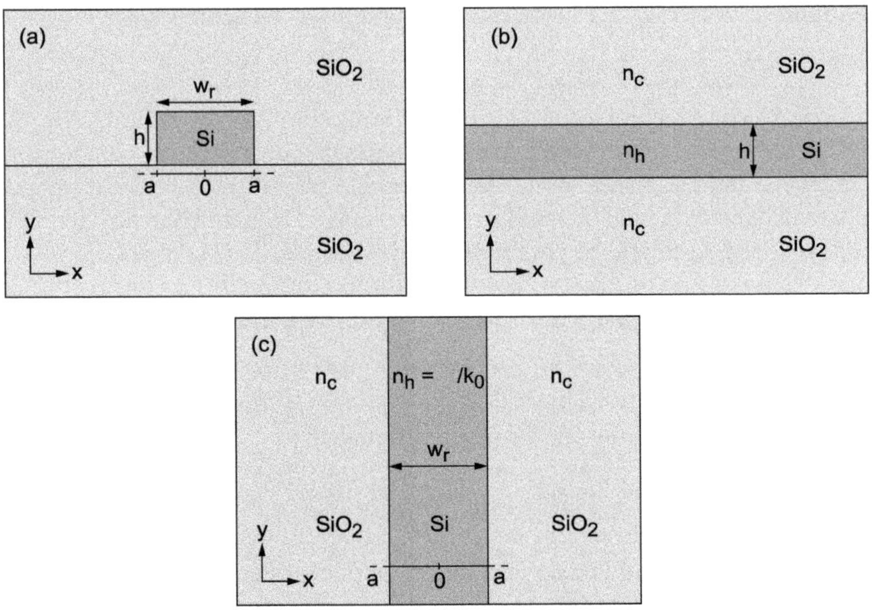

Fig. 2.1. (a) Schematics of rectangular optical waveguide, (b) Equivalent rectangular waveguide along x direction, and (c) Equivalent rectangular waveguide along y direction

Characteristics such as modal field distribution and effective index of the complex composite waveguide structure can be investigated through infinite slab waveguide system. This approximation method is called as Effective Index Method (EIM) (Pollock and Lipson, 2003a). EIM consider any arbitrary three dimensional waveguides as superposition of two orthogonal infinite planar waveguides. This method is useful in obtaining propagation characteristics of an optical waveguide

(Pollock and Lipson, 2003a). The cross section of the rectangular waveguide shown in Fig. 2.1(a) is divided into two orthogonal planar waveguides as shown in Fig. 2.1(b) and (c). The fundamental TE modes will have electric field (E_y) and magnetic field (H_x) components for waveguide depicted in Fig. 2.1(b). The effective index (β'/k_0) obtained by solving transcendental equation pertaining to the waveguide in Fig. 2.1(b) is substituted for refractive indices of the silicon waveguide in Fig. 2.1(c) which is infinite along y axis. Thus, the electric field (E_x) and magnetic field (H_y) in this waveguide are components of fundamental TM mode (Almeida *et al.*, 2004). Using similar analysis for this waveguide, a characteristic equation is obtained that gives the propagation constant and modal field distribution for the rectangular waveguide.

The refractive index distribution of planar slab waveguide shown in Fig. 2.1(b) is n_c for SiO_2 region and n_h for Si region. The boundary conditions are then applied, which is H_y and E_z need to be continuous. The characteristic Eigen value equation is obtained as

$$\tan(\kappa' h) = 2\left(\frac{\gamma'}{\kappa'}\right) \Big/ \left(1-\left(\frac{\gamma'}{\kappa'}\right)^2\right)$$

Here, $\gamma' = \sqrt{k_0^2 n_c^2 - \beta'^2}$ and $\kappa' = \sqrt{k_0^2 n_h^2 - \beta'^2}$ are the transverse component of field wavenumber in the SiO_2 and Si region. After solving the Eigen value equation, the value of effective index n'_h (= β'/k_0) of the slab waveguide structure of Fig. 2.1(b). Further, obtained effective index (n'_h) is substituted in place of the silicon waveguide refractive (n_h) in Fig. 2.1(c) which will give effective index of the composite structure.

Now, in the waveguide region of Fig. 2.1(c), Eq. 2.13 will be,

$$\frac{\partial^2 H_y}{\partial x^2} + \kappa^2 H_y = 0 = 0 \qquad ...(2.15)$$

where, $\kappa = \sqrt{k_0^2 n_h'^2 - \beta^2}$, n'_h is the refractive index of the waveguide region in Fig. 2.1(c). Similarly, in the cladding region, Eq. 2.13 will be:

$$\frac{\partial^2 H_y}{\partial x^2} + \gamma^2 H_y = 0 \qquad ...(2.16)$$

where, $\gamma = \sqrt{k_0^2 n_c^2 - \beta^2}$, n_c is the refractive index of the cladding region. Corresponding transcendental equation is given by:

$$\tan(\kappa a) = \left(\frac{n'_h}{n_c}\right)^2 \left(\frac{\gamma}{\kappa}\right) \qquad ...(2.17)$$

The general solution of Eq. 2.15 in the waveguide region will be

$$H_y = A\cos(\kappa x) \qquad |x| \le a \qquad ...(2.18)$$

Similarly, the solution of Eq. 2.16 in the cladding region will be

$$H_y = \begin{cases} Be^{-\gamma(x-a)} & x \ge a \\ Ce^{+\gamma(x+a)} & x \ge -a \end{cases} \qquad ...(2.19)$$

where, the constant terms A, B, and C can be calculated using boundary conditions at $x = \pm a$. At $x = \pm a$, H_y and E_z should be continuous. The expression of E_x and E_z in term of H_y is

$$E_x = \frac{\beta}{\omega\varepsilon_0\varepsilon_r} H_y \qquad ...(2.20)$$

$$E_z = \frac{1}{j\omega\varepsilon_0\varepsilon_r}\frac{\partial H_y}{\partial x} \qquad ...(2.21)$$

The expression of A, B and C after applying boundary condition will be

$$B = C = A\left(\frac{n_h'}{n_c}\right)^2\left(\frac{\gamma}{\kappa}\right)\sin(\kappa a) \qquad ...(2.22)$$

Substituting the values of B and C, region wise magnetic and electric fields can be written as:

$$H_y = \begin{cases} A\left(\frac{\beta}{\omega\varepsilon_0}\right)\left(\frac{n_c}{n_h'}\right)^2\left(\frac{\kappa}{\gamma}\right)\sin(\kappa a)e^{\gamma(x+a)} & x < -a \\ A\left(\frac{\beta}{\omega\varepsilon_0}\right)\cos(\kappa x) & |x| < a \\ A\left(\frac{\beta}{\omega\varepsilon_0}\right)\left(\frac{n_c}{n_h'}\right)^2\left(\frac{\kappa}{\gamma}\right)\sin(\kappa a)e^{-\gamma(x-a)} & x > a \end{cases} \qquad ...(2.23)$$

$$E_x = \begin{cases} A\left(\frac{\beta}{\omega\varepsilon_0}\right)\left(\frac{\kappa}{\gamma}\right)\sin(\kappa a)e^{\gamma(x+a)} & x < -a \\ A\left(\frac{\beta}{\omega\varepsilon_0}\right)\cos(\kappa x) & |x| < a \\ A\left(\frac{\beta}{\omega\varepsilon_0}\right)\left(\frac{\kappa}{\gamma}\right)\sin(\kappa a)e^{-\gamma(x-a)} & x > a \end{cases} \qquad ...(2.24)$$

Corresponding normalized field of Eq. 2.23 and Eq. 2.24 are as follows:

$$H_y = \begin{cases} \left(\frac{n_c}{n'_h}\right)^2 \left(\frac{\kappa}{\gamma}\right) \sin(\kappa a) e^{\gamma(x+a)} & x < -a \\ \cos(\kappa x) & |x| < a \\ \left(\frac{n_c}{n'_h}\right)^2 \left(\frac{\kappa}{\gamma}\right) \sin(\kappa a) e^{-\gamma(x-a)} & x > a \end{cases} \quad ...(2.25)$$

$$E_x = \begin{cases} \left(\frac{\kappa}{\gamma}\right) \sin(\kappa a) e^{\gamma(x+a)} & x < -a \\ \cos(\kappa x) & |x| < a \\ \left(\frac{\kappa}{\gamma}\right) \sin(\kappa a) e^{-\gamma(x-a)} & x > a \end{cases} \quad ...(2.26)$$

The average poynting vector can be written as:

$$P_{avg} = \frac{1}{2}\left|\vec{E} \times \vec{H}\right| \quad (2.27)$$

2.5 FUNDAMENTAL AND HIGHER ORDER MODES

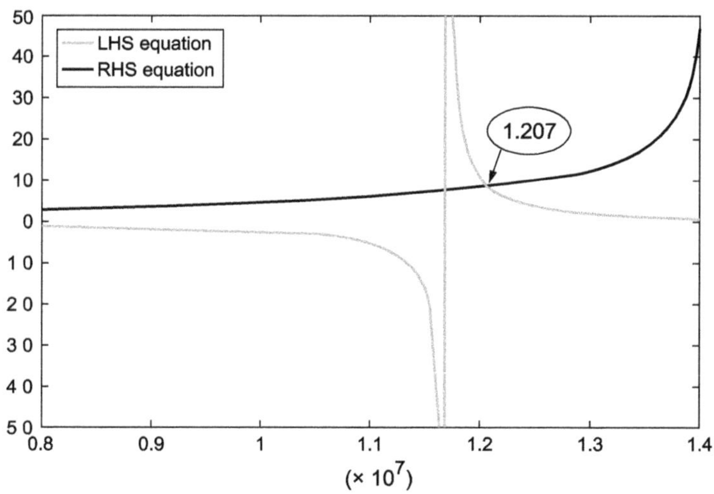

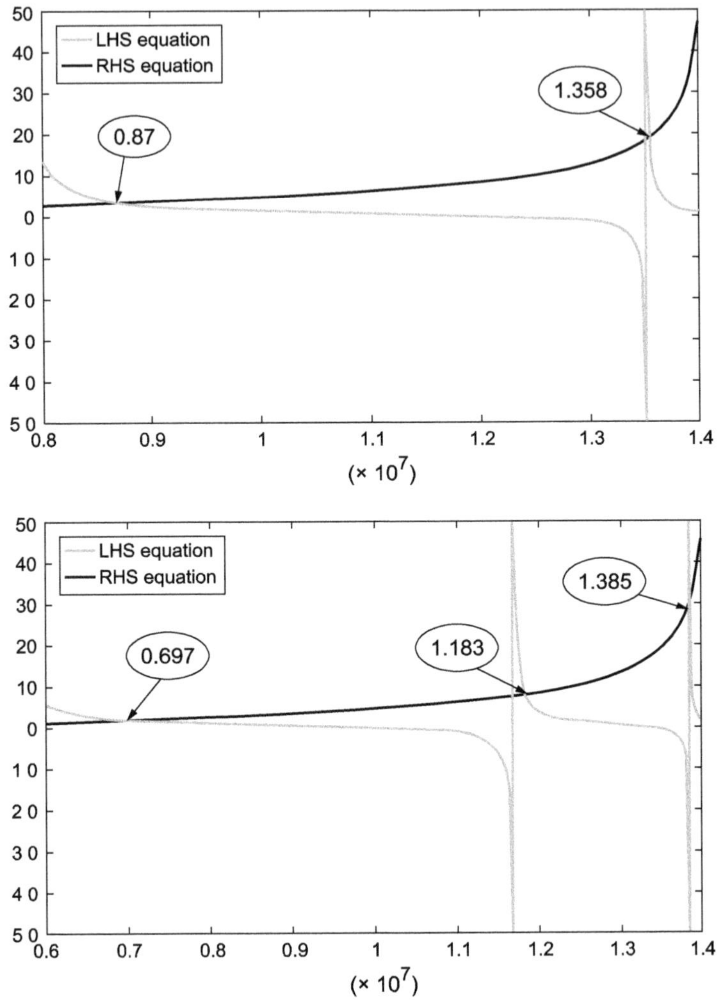

Fig. 2.2. Solution for β from Eq. 2.17, when w_r is (a) 400 nm, (b) 800 nm, and (c) 1200 nm.

The rectangular optical waveguide for SOI based PIC technology is theoretically derived so far. Hereafter, TM mode will be analyzed for three different waveguide widths. For illustrative purpose, $w_r = 400$ nm, 800 nm and 1200 nm is considered. The conditions for first and higher order modes are analyzed mathematically and justified with the FEM based simulation. The plot of left hand-side and right hand-side of Eq. 2.17 gives the point of intersection, i.e., the solution for beta (β). For waveguide widths of 400 nm, 800 nm and 1200 nm, corresponding solutions for β are shown in Fig. 2.2.

The value of β for highest order supportive mode in 400 nm, 800 nm and 1200 nm waveguide widths are 1.207×10^7 rad/m, 1.358×10^7 rad/m and 1.385×10^7

rad/m, respectively. The corresponding mode structure in *x-y* cut plane for these three w_r are shown in Fig. 2.3.

Fig. 2.3. Solution for highest order mode when w_r is (a) 400 nm, (b) 800 nm, and (c) 1200 nm.

As one of the main objectives is to compare and verify theoretical mode field profile with FEM based field profile, for which waveguide width of 400 nm with its corresponding theoretical and simulated FEM based normalized electric field (E_x), normalized magnetic field (H_y) and normalized power (P_{avg}) is shown below.

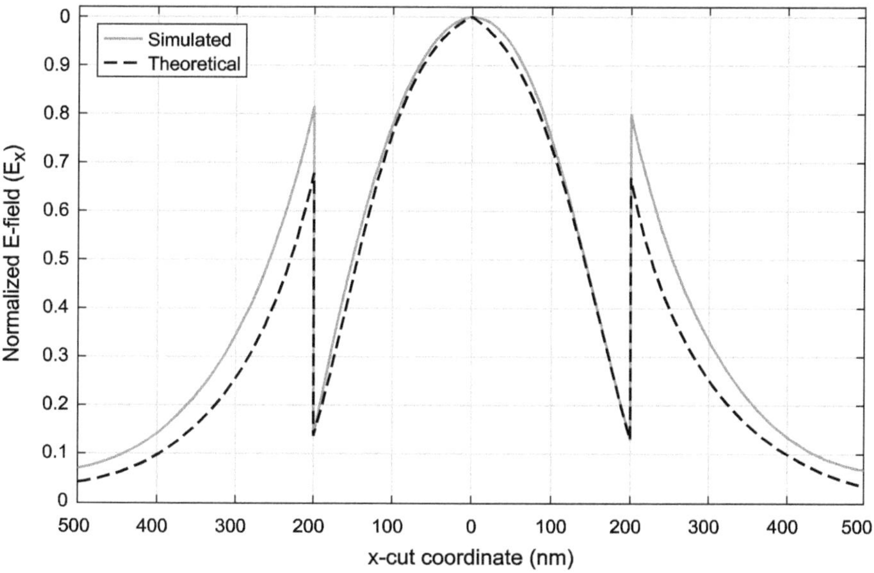

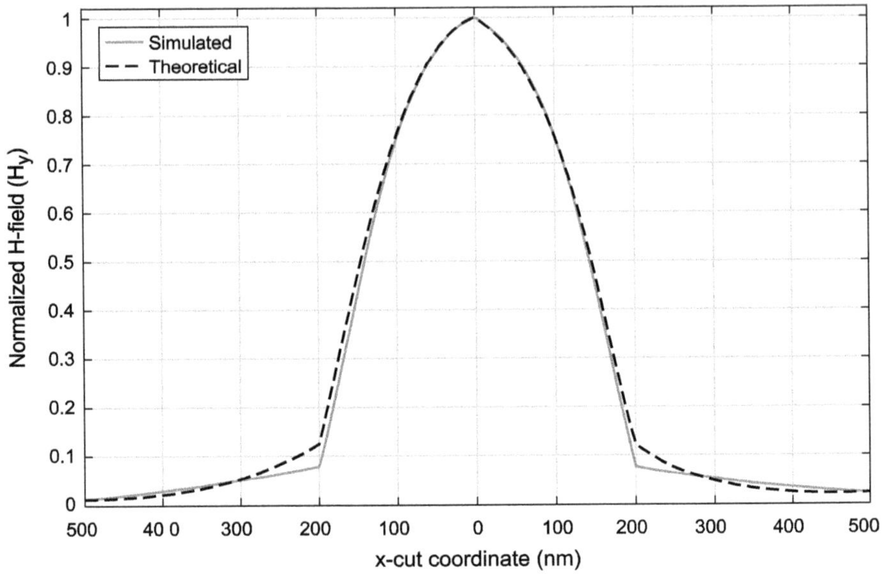

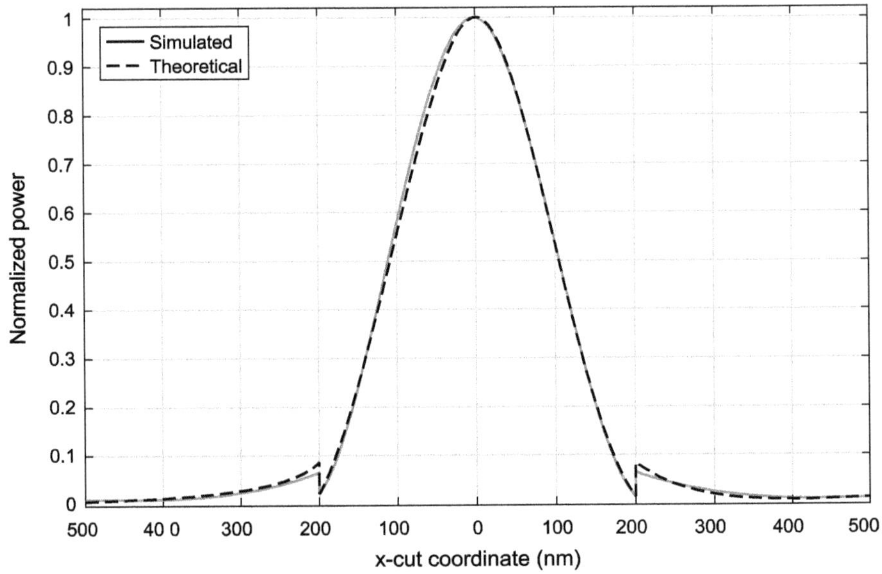

Fig. 2.4. For w_r of 400 nm: (a) Normalized E_x, (b) Normalized H_y, and (c) Normalized power.

From Fig. 2.4, it is evident that most of the normalized electric and magnetic fields as well as the normalized power are confined to the guiding region of the waveguide. Also, it is evident that the waveguide will support only the first order mode for the 400 nm waveguide dimension which is justified by both the theoretical and simulated results. Furthermore, for $w_r = 800$ nm, normalized intensity profile of first and second order modes are shown in Fig. 2.5. From this figure, it is clear that the rectangular buried waveguide will support first order as well as second order modes. This is also evident theoretically by the two solutions of β from Fig. 2.2(b).

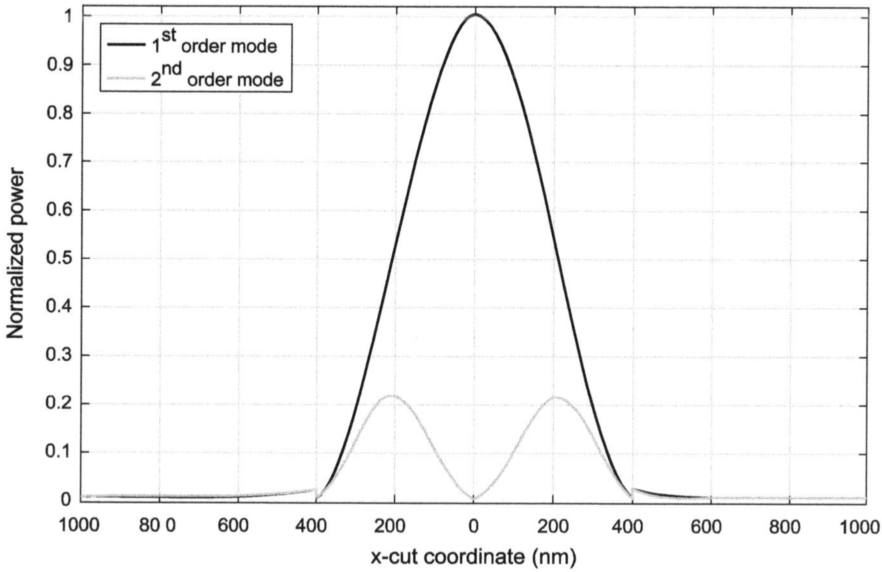

Fig. 2.5. Normalized power profile for different modes of w_r = 800 nm.

Again, if the waveguide width is increased to 1200 nm, it will support third order mode as well, whose corresponding absolute normalized power is shown below in Fig. 2.6.

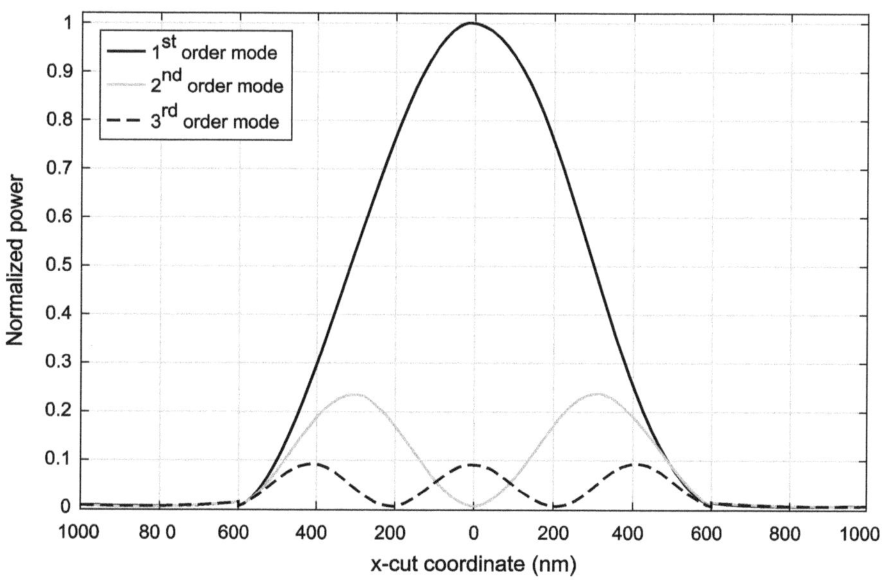

Fig. 2.6. Normalized power profile for different modes of w_r = 1200 nm.

The mode equations can be derived for buried rectangular waveguide having waveguide total width compatible for guiding more than one fundamental mode. This explain the dependency of higher order modes on waveguide parameters with the help of both theoretical and analytical methods. The mode profile of rectangular waveguide is shown graphically in Fig. 2.6, where guiding inside higher refractive indexed material is achieved for 1200 nm waveguide width.

❑❑❑

3 Theoretical Analysis of Slot Waveguide

This chapter is organized to gives a brief review of theoretical modal investigation of generalized field equations for various slot waveguides. Initially, a single slot based structures is briefly discussed having three types of slot waveguides with configuration of vertical, horizontal and cross slot, where guiding phenomena occurs in low refractive indexed material. These waveguides are examined with respect to waveguide parameters which are further optimized with the help of confinement factor and effective mode area. A theoretical analysis is carried out for all the three slot waveguides to plot the behavior of parameters i.e. electric and magnetic fields along with poynting vectors. The same parameters have been justified with the help of Finite Element Method (FEM) technique. This analytical approach gives insight into the characteristic features of linear slot waveguides. Although, the quantitative effectiveness of effective refractive index (β/k_0) calculated from theoretical analysis is slightly underestimated as compared to FEM results (Almeida *et al.*, 2004; Priye, Malviya and Mickelson, 2018). Further in this chapter, multiple slot waveguide based on vertically configured single slot waveguide is thoroughly investigated by extending it to three and five slots, respectively. The odd number of slots are studied because of the centro symmetricity in *y*-axis and presence of slot at pivot. Theoretically derived results are compared with FEM results and shown graphically for all these multiple slot conditions.

3.1 SINGLE SLOT WAVEGUIDE

In a slot waveguide structure, a low indexed slot of few nanometers (i.e., ~100 nm) width is sandwiched between two high indexed slabs or bricks. Low refractive indexed material is taken as SiO_2 and high refractive index material is taken as Si, which is used here to construct suitable waveguide for SOI based PICs. The refractive index of SiO_2 and Si are taken as 1.444 and 3.4752, respectively at communication wavelength of 1550 nm. In slot waveguide, light will dominantly propagate through the region of low indexed slot. For this to happen, waveguide structural parameters play a crucial role in achieving efficient guiding properties. The theoretical discussion and realization of slot waveguide is divided in two parts

DOI: 10.1201/9781003783589-3

i.e., single slot waveguide and multiple slot waveguide. In the beginning of this slot waveguide study, the case of single slot waveguide is elaborated.

Depending upon the geometry of single slot waveguide structure, optical slot waveguides can be broadly classified into three categories; vertical, horizontal and cross-slot waveguides. Till date, it has been reported that vertical and horizontal slot waveguides have more applications than that of cross-slot (Dar *et al.*, 2012; Azuelos *et al.*, 2017). Both quasi-TE and quasi-TM modes can be supported by vertical and horizontal slot waveguides. Moreover, quasi-TE mode confines better in the slot region in case of vertical slot waveguide and similarly, quasi-TM mode does in horizontal slot waveguide (Xu *et al.*, 2004). Silicon-On-Insulator (SOI) technology chosen for fabricating these slot waveguides is of 220 nm height (Xu *et al.*, 2004; Dar *et al.*, 2012; Viphavakit *et al.*, 2015). In vertical slot, the fabrication process involves etching of narrow slot region, which results in increased scattering loss (Sun *et al.*, 2007). However, this loss is not prominent in case of fabricating horizontal slot waveguide because it can be fabricated using Chemical Vapor Deposition (CVD) and thermal oxidation techniques (Kumari, Varshney and Pal, 2018). In horizontal slot, width can be controlled more precisely, and thinner slot can be formed with smoother surfaces (Barrios and Lipson, 2005). These techniques can be applied layer by layer, which leads to lesser surface roughness. So, the scattering loss is reduced significantly in horizontal slot as compared to vertical slot (Sun *et al.*, 2007; Kumari, Varshney and Pal, 2018). The cross-slot waveguide has horizontal as well as vertical slot which are orthogonal to each other and can be fabricated by Plasma Enhanced Chemical Vapor Deposition (PECVD). In PECVD, lithography is followed for horizontal slot layer and dry etching is performed for the vertical slot and silicon rails (Rukhlenko, Premaratne and Agrawal, 2012). The cross-slot waveguide supports both quasi-TE and quasi-TM modes independently, for two different values of propagation constant (β). In the initial stage, detailed investigation is performed below on parametric optimization of vertical slot waveguide.

3.2 VERTICAL SLOT

In general, mode analysis approach is used at cross-section of the waveguide to compute the propagation constant, wave number and mode shapes of propagating mode inside a waveguide. However, to solve the mode properties, rigorous Finite Element Method (FEM) (Arti Agrawal, 2013) based on time-harmonic electromagnetic field distribution is used. The optimization of physical parameters of waveguides is achieved by: (*i*) maximizing the slot confinement factor (Γ_s), and (*ii*) minimizing the effective mode area (A_{eff}). The slot confinement factor (CF) and the effective mode area (Khanna *et al.*, 2009) are given in the following equations as:

$$\Gamma_s = \frac{\iint_{\text{slot}} |S(x,y)| dxdy}{\iint_{\text{total}} |S(x,y)| dxdy} \qquad ...(3.1)$$

$$A_{eff} = A_{eff} = \frac{\left(\iint_{total} |S(x,y)| dxdy \right)^2}{\iint_{slot} |S(x,y)|^2 dxdy} \quad ...(3.2)$$

Where, $S(x, y)$ is the poynting vector profile of dominant first order mode, which can be expressed as $|S(x, y)| = |E(x, y) \times H(x, y)|$. Where, $E(x, y)$ and $H(x, y)$ are transverse electric and magnetic field profiles of mode distribution, respectively.

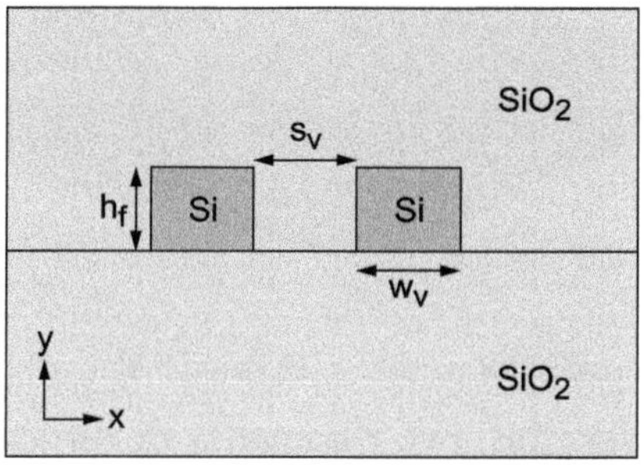

Fig. 3.1. Schematics of vertical slot waveguide.

In the case of vertical slot waveguide, as shown in Fig. 3.1, the dominant component of electric and magnetic fields are in x and y direction, respectively, i.e., $E(x, y) = E_x$ and $H(x, y) = H_y$.

3.2.1 Optimization of Waveguide Parameters

Waveguide structural parameters play a major role in the distribution of electric and magnetic field components inside the guided region, which result in the variation

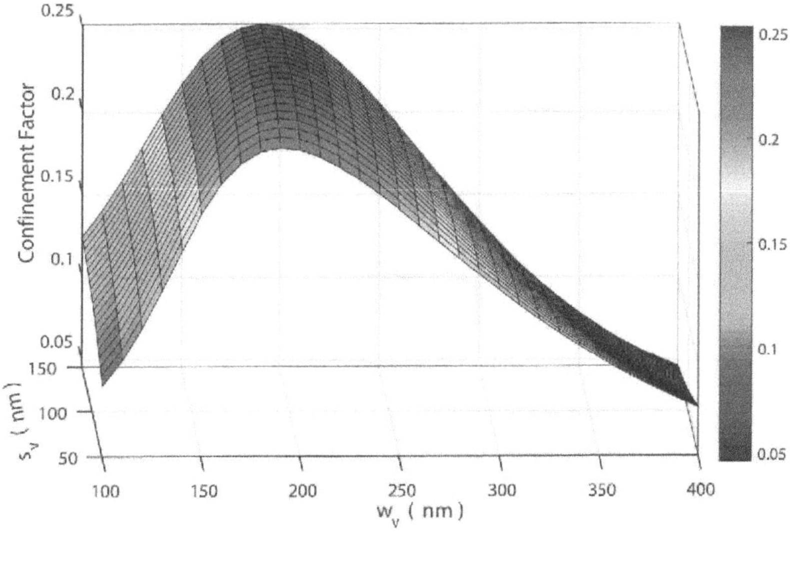

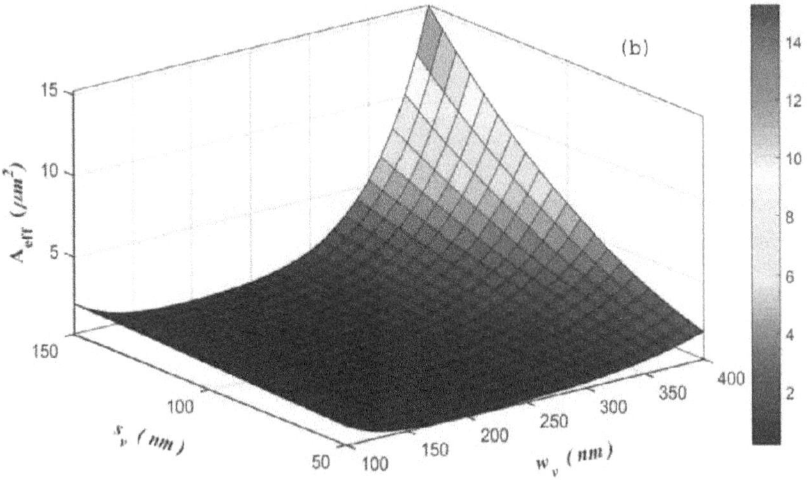

Fig. 3.2. Optimization of w_v and s_v in vertical slot: (a) Γ_s; and (b) A_{eff}.

of Γ_s and A_{eff}. To illustrate this, two parameters of vertical slot waveguide: s_v is varied from 50 nm to 150 nm and w_v from 100 nm to 400 nm. While varying s_v and w_v, the third parameter h_f is fixed to 220 nm as per SOI standard fabrication technology. The value of electric and magnetic field propagating inside mode structure is obtained using Finite Element Method, from which the values of Γ_s and A_{eff} is further calculated using Eq. 3.1 and Eq. 3.2, respectively. The corresponding value of Γ_s and A_{eff} for various combinations of s_v and w_v is shown in Fig. 3.2.

The higher value of Γ_s is shown with red color in sidebar of the Fig. 3.2(a) and lower value of A_{eff} illustrated with blue color in Fig. 3.2(b), which are desired values for optimizing vertical slot waveguide. The optimized value of s_v and w_v

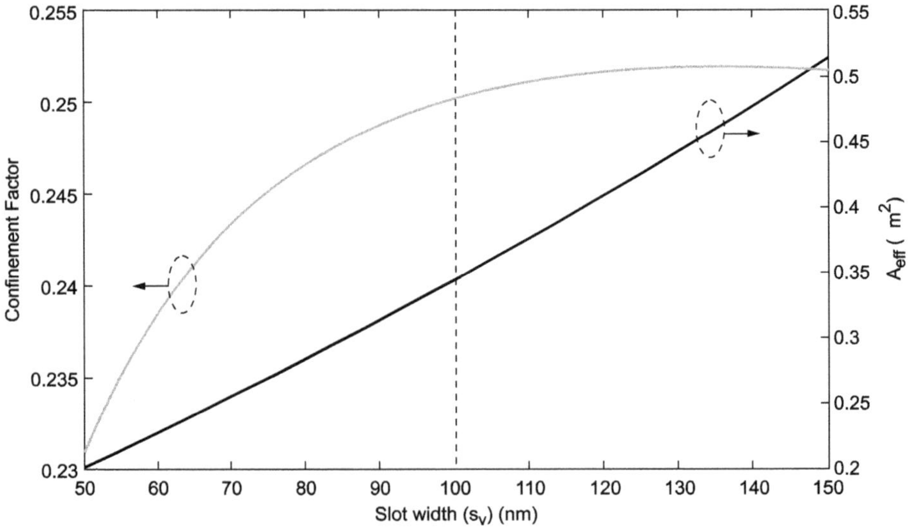

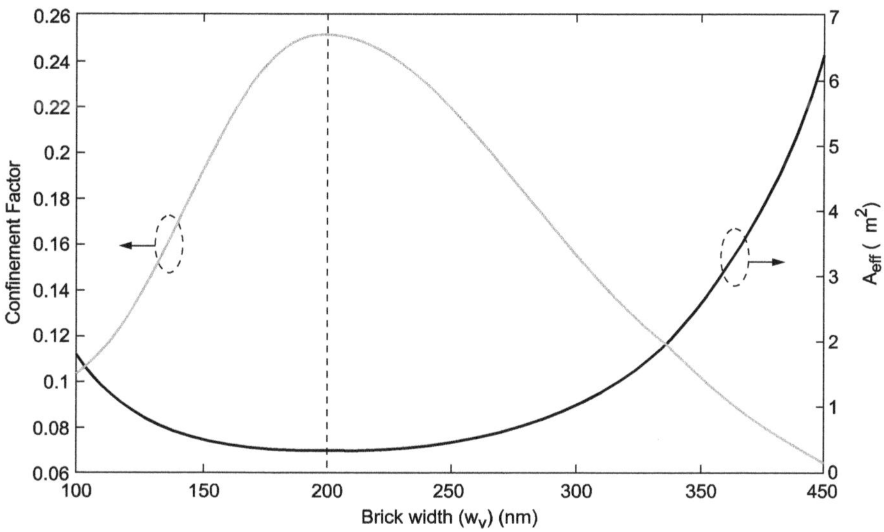

Fig. 3.3. Optimization of (a) s_v and (b) w_v; for maximum Γ_s and minimum A_{eff} in vertical slot.

corresponding to the highest value of Γ_s and the lowest value of A_{eff} are 100 nm and 200 nm, respectively. To attain the optimized value, the profile of s_v is first observed by keeping w_v constant, which is plotted in Fig. 3.3(a). The value of Γ_s becomes constant after certain value of slot width i.e. 100 nm, for which the w_v profile is further observed by keeping s_v constant, as shown in Fig 3.3(b). In the first case, a parametric sweep from 50 nm to 150 nm is applied to s_v, as the value of s_v is increased, A_{eff} is also increased, but the value of Γ_s get saturated to 25.02% at s_v = 100 nm. However, in the second case, the value of s_v is fixed to 100 nm and a sweep from 100 nm to 400 nm is applied to w_v. Results show that the value of A_{eff} is minimum and Γ_s is maximum at brick width of 200 nm. From the above analysis, it is inferred that the optimized value of s_v and w_v is 100 nm and 200 nm, respectively.

3.2.2 Mode in Vertical Slot Waveguide

As discussed in the previous section, x-component of electric field and y-component of magnetic field are the only dominant field profiles for vertical slot waveguide. Thus, in vertical slot waveguide, the quasi-TE mode is the only dominant mode in the guiding vertical slot region. The TM mode from Eq. 2.12 is revised for cover, brick and slot region as:

$$\frac{\partial^2 H_y}{\partial x^2} - \gamma_c^2 H_y = 0 \quad ...(3.3)$$

$$\frac{\partial^2 H_y}{\partial x^2} + \kappa_h^2 H_y = 0 \quad ...(3.4)$$

$$\frac{\partial^2 H_y}{\partial x^2} - \gamma_s^2 H_y = 0 \qquad ...(3.5)$$

Where, $\gamma_c = \sqrt{\beta^2 - k_0^2 n_c^2}$, $\kappa_h = \sqrt{k_0^2 n_h^2 - \beta^2}$, and $\gamma_s = \sqrt{\beta^2 - k_0^2 n_s^2}$. Here, n_c, n_h, n_s are the refractive index of cover, brick, and slot region, respectively. κ_h is the transverse wave number in high-indexed bricks, γ_c is the field decay coefficient in cladding, and γ_s is the field decay coefficient in vertical slot (Almeida *et al.*, 2004). The γ_c, κ_h, and γ_s follow the combined relation as:

$$\beta^2 = k_0^2 n_c^2 + \gamma_c^2 = k_0^2 n_h^2 - \kappa_h^2 = k_0^2 n_s^2 + \gamma_s^2 \qquad ...(3.6)$$

Where, k_0 is the free space wave number and β is the eigen-mode propagation constant, which can be obtained by solving the transcendental equation derived as:

$$\tan[\kappa_h(b-a)-\phi] = \left(\frac{n_h}{n_s}\right)^2 \left(\frac{\gamma_s}{\kappa_h}\right) \tan(\gamma_s a) \qquad ...(3.7)$$

Where, $\phi = \tan^{-1}\left[\left(\frac{n_h}{n_c}\right)^2 \left(\frac{\gamma_c}{\kappa_h}\right)\right]$.

The value of parameters *a* and *b* are designated to inner and outer boundary of each vertical brick due to the widely accepted symmetric geometry of vertical slot waveguide along *y*-axis. So, the regions of slot is expressed as $|x| < a$; brick is expressed as $a < |x| < b$; and side covers are expressed as $|x| > b$, which can be seen in Fig 3.5. From Eq. 3.7, the solution of transcendental equation is plotted in Fig. 3.4.

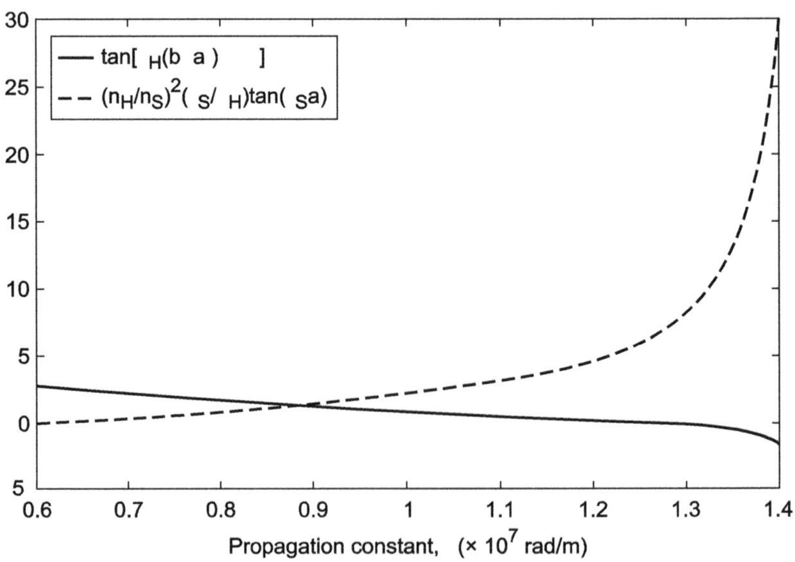

Fig. 3.4. Calculation of β for quasi-TE mode in vertical slot waveguide.

The derivation of mode field equations is challenging in this case as the interaction between field intensity is highly dependent on overlapped mode region generated from both the Si bricks. The constructive interference evolved from the evanescent field of Si bricks, inside slot region will result in low RI guiding waveguide. The resulting analytical solutions for the normalized transverse magnetic field is derived by assuming the solution of Eq. 3.3, Eq. 3.4, and Eq. 3.5 as:

$$H_y = \begin{cases} A\cosh(\gamma_s |x|) & |x| < -a \\ B\cos(\kappa_h |x-a|) + C\sin(\kappa_h |x-a|) & a < |x| < b \\ De^{-\gamma_c(|x|-b)} & |x| > b \end{cases} \quad ...(3.8)$$

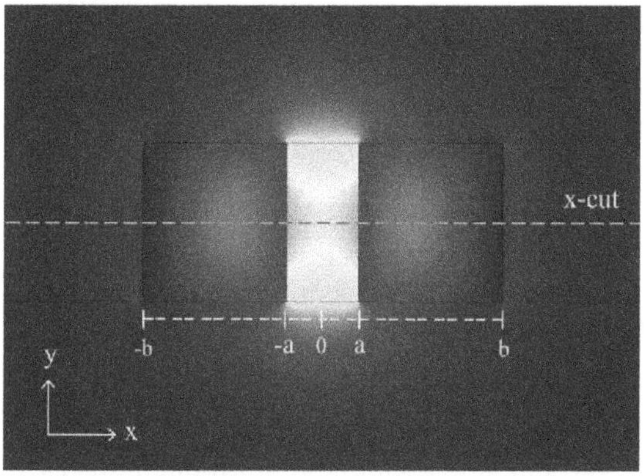

Fig. 3.5. Poynting vector profile in the slot region of vertical slot waveguide.

The expression of *B*, *C* and *D* is obtained after applying continuity of H_y and E_z boundary conditions using Eq. 2.21, which is

$$B = A\cosh(\gamma_s a)\, B = A\cosh(\gamma_s a) \quad ...(3.9)$$

$$C = A\left(\frac{n_h}{n_s}\right)^2\left(\frac{\gamma_s}{\kappa_h}\right)\sinh(\gamma_s a) \quad ...(3.10)$$

$$D = A[\cosh(\gamma_s a)\cos\{\kappa_h(b-a)\} + \left(\frac{n_h}{n_s}\right)^2\left(\frac{\gamma_s}{\kappa_h}\right)\sinh(\gamma_s a)\sin\{\kappa_h(b-a)\}] \quad (3.11)$$

By substituting the values of *B*, *C* and *D*, region wise to Eq. 3.8 and Eq. 2.20, magnetic and electric fields will be obtained. The above finding validates the working of vertical slot waveguide as a single mode waveguide. Since, E_x is horizontally polarized propagating in the slot region between both the bricks and perpendicular to the slot length, showing strong discontinuity at the boundary of bricks. Moreover, the intensity of electric field is found to be much higher inside the slot region. On the other hand, the magnetic field (H_y) is continuous at the along the length of slot, showing higher optical power intensity in the high-indexed region.

3.3 HORIZONTAL SLOT

The case of horizontal slot waveguide, shown in Fig. 3.6, has the dominant component of E and H in y and x direction, respectively. As a result, $E(x,y)$ will be E_y and $H(x,y)$ will be H_x. Using the approach used in optimization of vertical slot waveguide, the parameters of horizontal slot waveguide i.e., s_h and h_h are optimized.

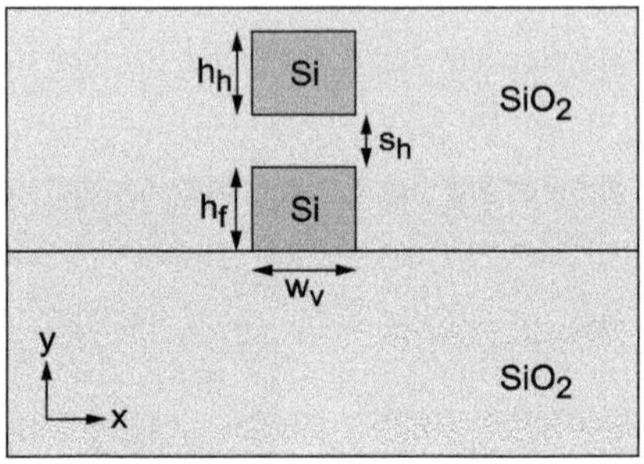

Fig. 3.6. Schematics of horizontal slot waveguide.

3.3.1 Optimization of Waveguide Parameters

During optimization process s_h and h_h are varied from 50 nm to 150 nm and 100 nm to 400 nm, respectively. Besides, the value of w_v is taken from previously optimized value of vertical slot waveguide, with the value of $h_f = 220$ nm as per standard height of the first silicon layer. The results of Γ_s and A_{eff} corresponding to s_h and h_h is plotted in the likewise approach as shown in Fig. 3.7(a) and 3.7(b), respectively.

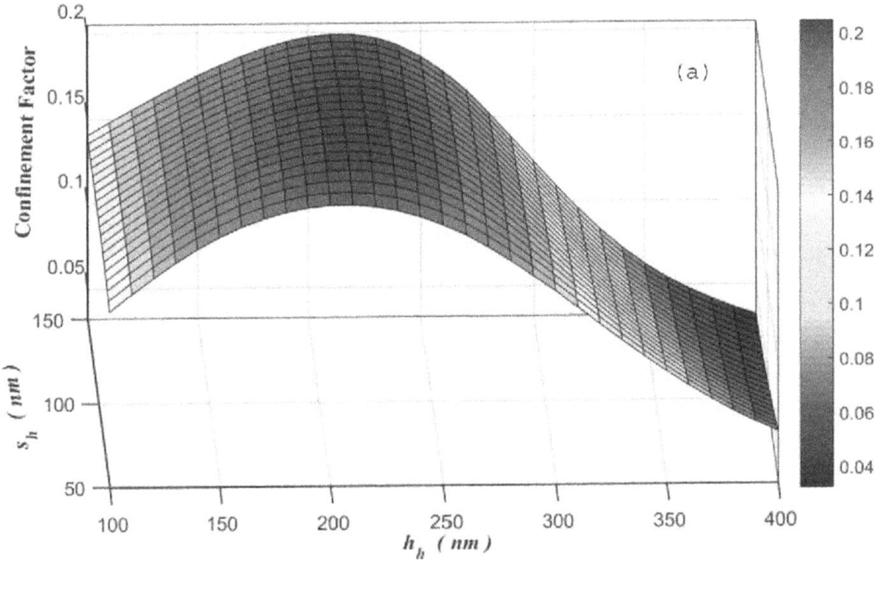

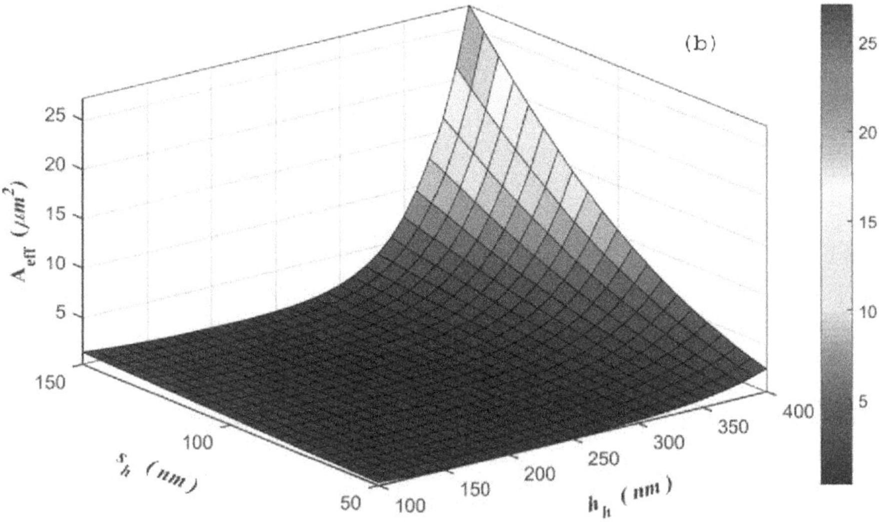

Fig. 3.7. Optimization of s_h and h_h in horizontal slot: (a) Γ_s; and (b) A_{eff}.

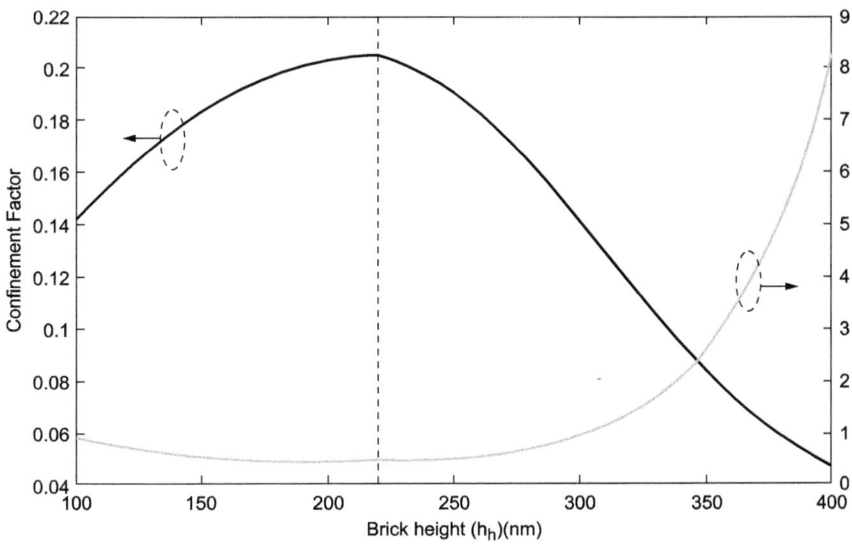

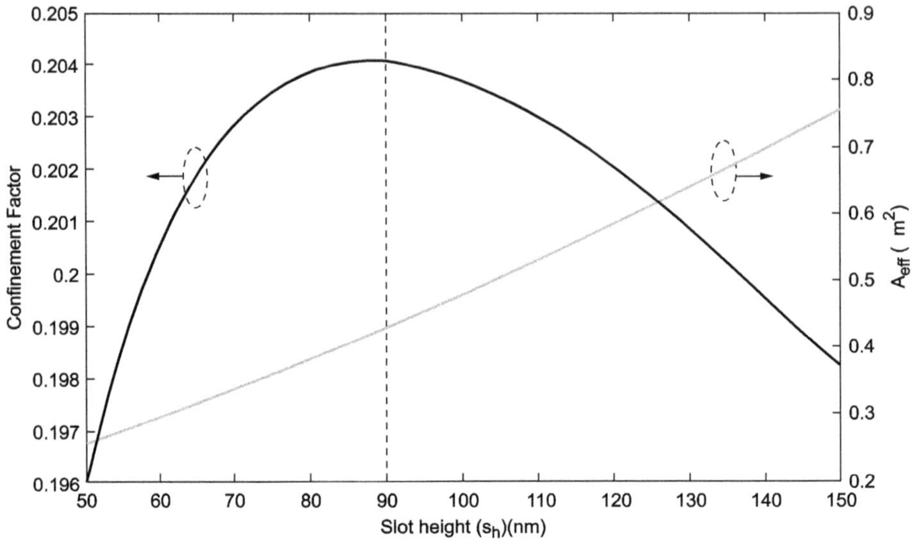

Fig. 3.8. Optimization of (a) h_h and (b) s_h for maximum Γ_s and minimum A_{eff} in horizontal slot.

3.3.2 Mode in Horizontal Slot Waveguide

The maximum confinement factor with minimum footprint of mode size is achieved at $s_h = 90$ nm and $h_h = 220$ nm. To validate this, individual analysis on h_h and s_h is performed and shown in Fig. 3.8(a) and 3.8(b), respectively. In the first

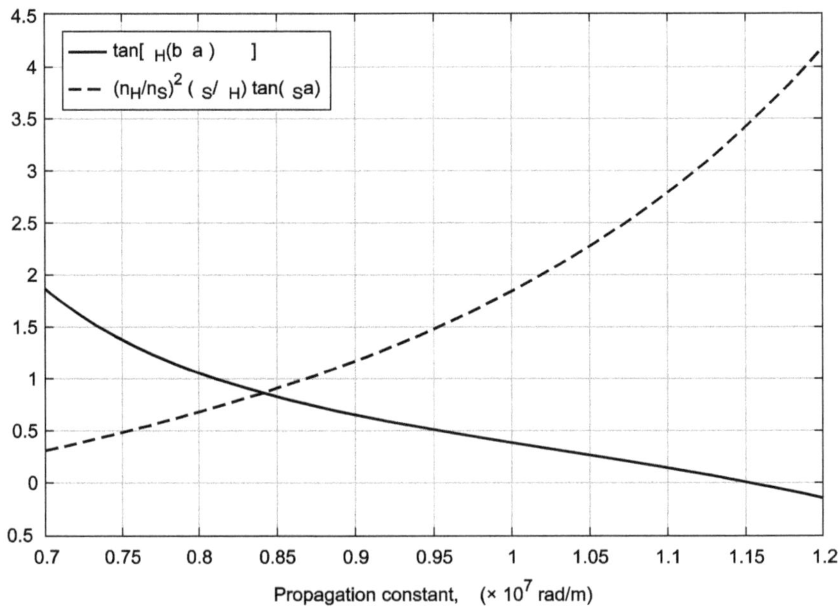

Fig. 3.9. Calculation of β for quasi-TM mode in horizontal slot waveguide.

step, parametric sweep is applied to h_h from 100 nm to 400 nm with resolution of 10 nm while keeping the s_h as fixed value to 90 nm. After this, sweep is applied to s_h from 50 nm to 150 nm with a resolution of 5 nm while keeping $h_h = 220$ nm. The criteria for maximum Γ_s and minimum A_{eff} can easily be shown using these results at the optimized value of $h_h = 220$ nm and $s_h = 90$ nm.

In case of horizontal slot waveguide, quasi-TM mode is the dominant mode in slot region (Xu *et al.*, 2004), where major component of electrical field is E_y. The calculation of β using transcendental Eq. 3.7 is been carried out for $a' = 45$ nm, $b' = 265$ nm (i.e., based on optimized parameters) at operating wavelength (λ) of 1550 nm, showing the corresponding values in Fig. 3.9.

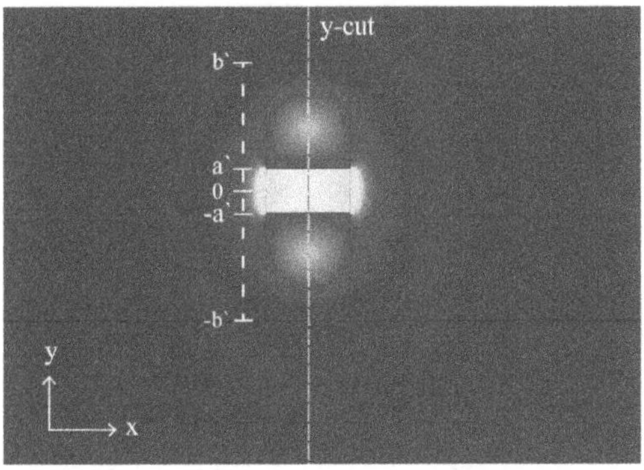

Fig. 3.10. Poynting vector profile in the slot region of vertical slot waveguide.

Fig. 3.9 shows that only one solution of β is confirmed for transcendental Eq. 3.7, under optimized parametric values of horizontal slot waveguide. In quasi-TM mode, transcendental equation will be same as Eq. 3.7. However, in mode field equation, E_x and H_y fields will be replaced by E_y and H_x, respectively, which will be an only function of y. After applying these modifications to Eq. 3.8 and Eq. 2.20, E_y is found to be perpendicular to the propagating length of the slot, showing strong discontinuity at the surface of slot-brick interface. Conversely, magnetic field (H_x) is found to be continuous at the slot-brick interface with much higher intensity in the low-indexed slot region as shown in Fig. 3.10.

3.4 CROSS SLOT

The third category of slot waveguide is shown in Fig. 3.11, which is a combination of horizontal and vertical slot waveguides. The findings from the analysis and optimization done in the previous two cases indicate that parametric values of horizontal and vertical waveguide can be efficiently used to

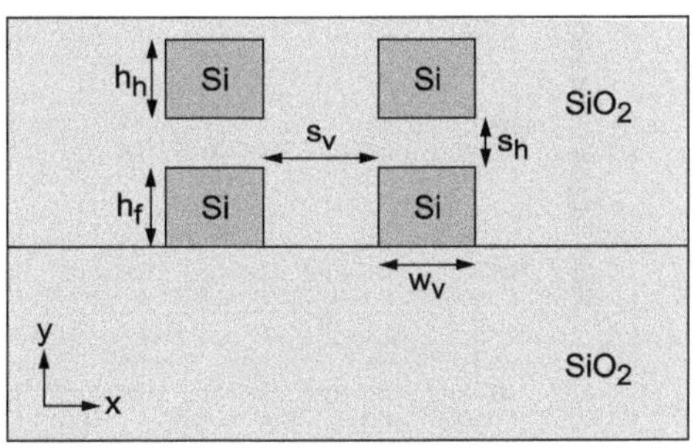

Fig. 3.11. Schematics of cross slot waveguide.

construct cross slot waveguide. A significant positive correlation between vertical and horizontal slot waveguides with the cross-slot waveguide can be found, if these waveguides are placed side-by-side.

In cross slot waveguide, two optimized horizontal slot waveguides are placed side-by-side leaving a gap of s_v, which is an optimized slot width in vertical slot waveguide as shown in Fig. 3.11. This waveguide is symmetric vertically as well as horizontally with respect to the common slot of combined vertical and horizontal slot waveguide.

3.4.1 Quasi-TE Mode in Cross Slot Waveguide

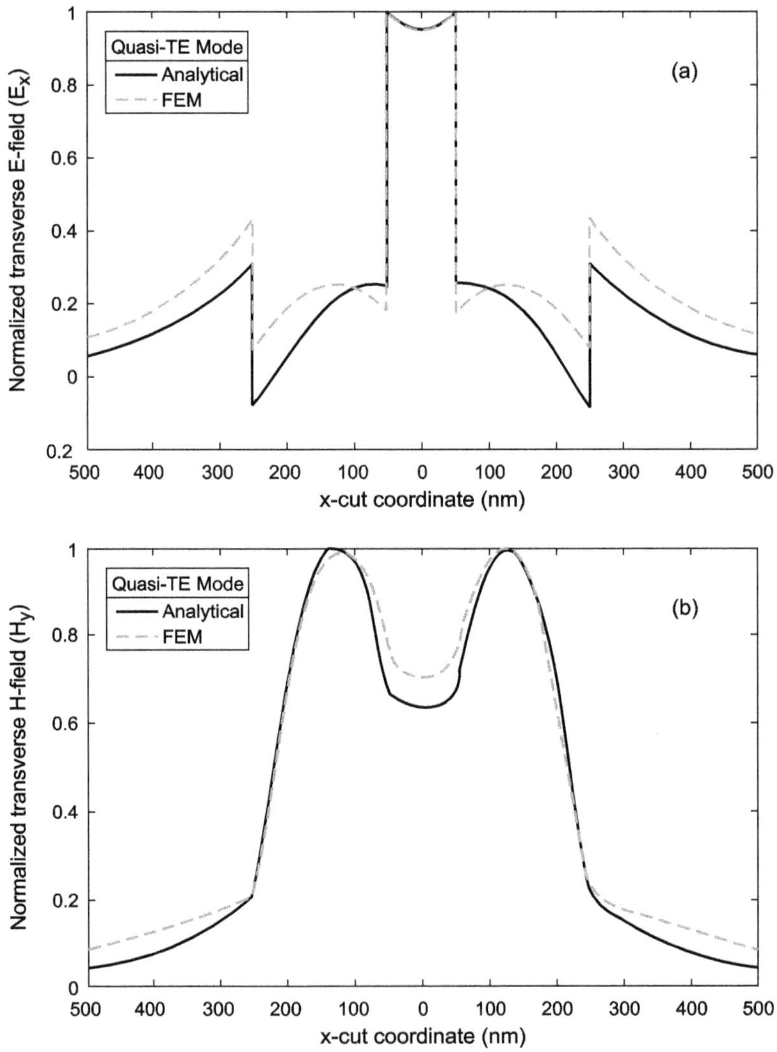

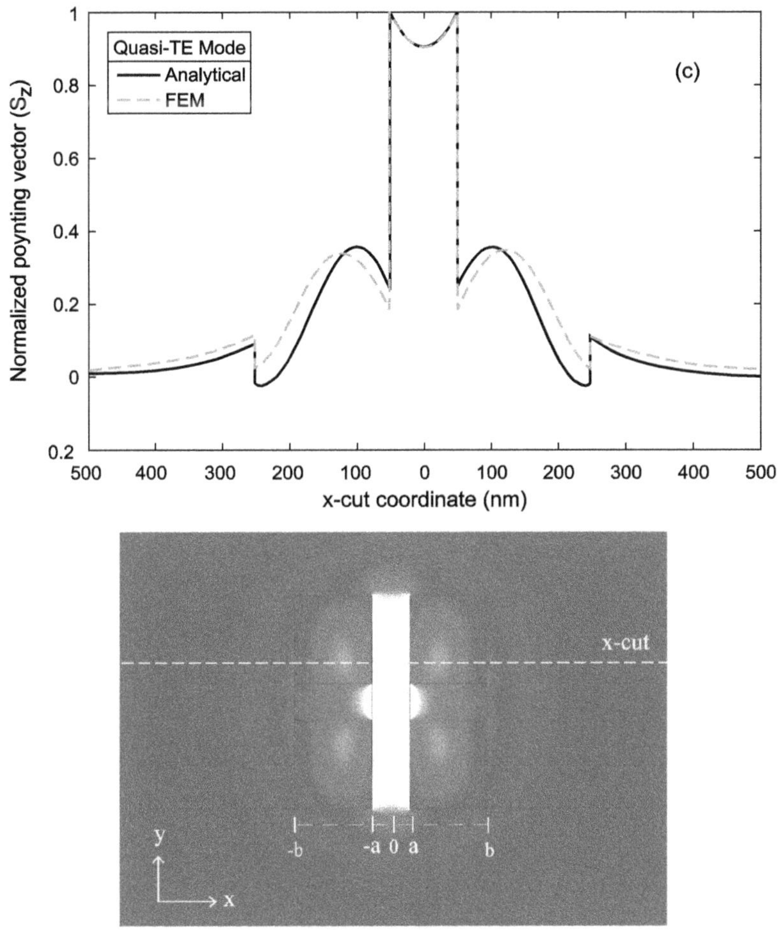

Fig. 3.12. Normalized field profiles of quasi-TE mode: (a) E_x; (b) H_y; (c) S_z and; (d) x-y plane poynting vector profile.

The modal properties of both vertical and horizontal slot waveguide is supported by this cross slot waveguide. This waveguide supports quasi-TE mode while working like a vertical slot waveguide, and quasi-TM mode while working like a horizontal slot waveguide. For quasi-TE mode, values of normalized electric field, magnetic field and optical power are evaluated from Eq. 2.20, 3.8 and 2.27, respectively. These analytical results are then compared to FEM based results as plotted in Fig. 3.12.

3.4.2 Quasi-TM Mode in Cross Slot Waveguide

In a similar way, the values of normalized electric field, magnetic field and optical power of quasi-TM mode are evaluated and compared with FEM based results as shown in Fig. 3.13.

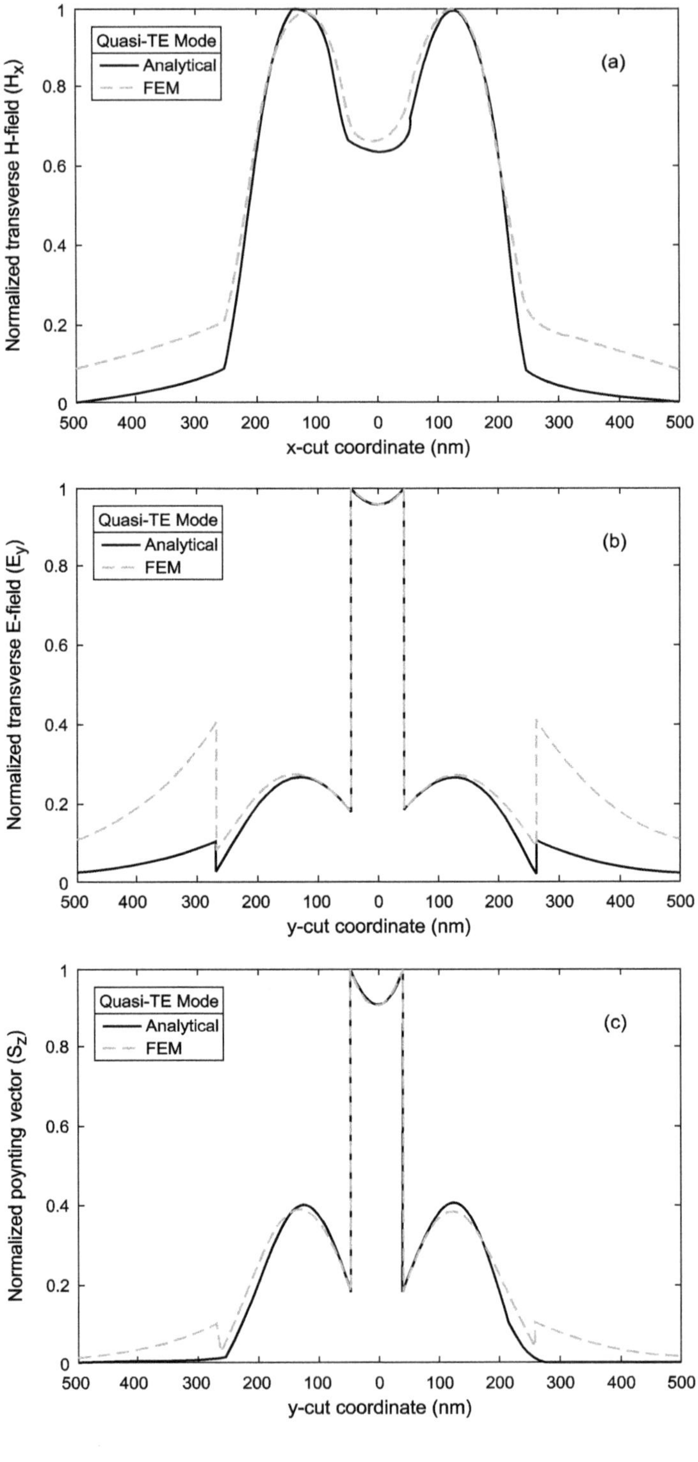
Quasi-TE Mode
Analytical
FEM
(a)
Normalized transverse H-field (H_x)
x-cut coordinate (nm)
500 400 300 200 100 0 100 200 300 400 500
0 0.2 0.4 0.6 0.8 1
Quasi-TE Mode
Analytical
FEM
(b)
Normalized transverse E-field (E_y)
y-cut coordinate (nm)
500 400 300 200 100 0 100 200 300 400 500
0 0.2 0.4 0.6 0.8 1
Quasi-TE Mode
Analytical
FEM
(c)
Normalized poynting vector (S_z)
y-cut coordinate (nm)
500 400 300 200 100 0 100 200 300 400 500
0 0.2 0.4 0.6 0.8 1

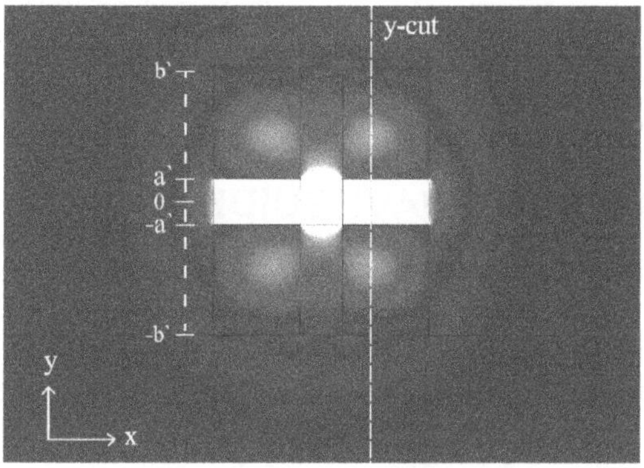

Fig. 3.13. Normalized field profiles of quasi-TM mode: (a) E_y; (b) H_x; (c) S_z and; (d) x-y plane poynting vector profile.

The cross-slot waveguide is a hybrid structure of vertical and horizontal slot waveguide, which supports both quasi-TE and quasi-TM modes for two entirely different values of β. The values of β for these quasi-TE and quasi-TM modes are calculated from Eq. 3.7. In the case of quasi-TE dominant mode, the calculated value of β is 0.881447×10^7 rad/m for which most of the energy is confined in vertical slot region as shown in Fig. 3.12(d). However, in quasi-TM dominant mode, the value of β is calculated as 0.842640×10^7 rad/m which supports the confinement of energy in horizontal slot region as shown in Fig. 3.13(d). Thus, the corresponding value of n_{eff} for guiding inside the vertical slot and horizontal slot structure are 2.1744 and 2.0787, respectively. The two values of n_{eff} will produce a birefringence between two modes. The value of birefringence is calculated as 3.8807×10^5 rad/m, which shows strong-birefringence, which can support both the modes simultaneously inside cross-slot waveguide.

Table 3.1. Optimized data for single slot waveguide

Parameter	Value	Aeff (μm^2)	Γ_s (%)
	Vertical slot waveguide		
Slot width (*sv*)	100 nm	0.346	25.02
Slot height (*hf*)	220 nm		
Brick width (*wv*)	200 nm		
	Horizontal slot waveguide		
Upper brick height (*hh*)	220 nm	0.428	20.41
Lower brick Height (*hf*)	220 nm		
Slot height (*sh*)	90 nm		
Brick width (*wv*)	200 nm		
	Cross-slot waveguide		
Slot width (*sv*)	100 nm	**Quasi-TE**	
Slot height (*sh*)	90 nm	0.438	34.16

Brick width (*wv*)	200 nm	**Quasi-TM**	
Upper brick height (*hh*)	220 nm	0.480	30.12
Lower brick Height (*hf*)	220 nm		

3.5 MULTIPLE SLOT WAVEGUIDE

As discussed in earlier sections, vertical slot waveguide has TM mode guidance inside slot region. Since the component of electric field (E_z) vanishes during the derivation of Helmholtz equation for complex structure, this mode is known to be quasi-TE mode that have dominant optical guidance in slot region.

3.5.1 Mode in 'One' Slot Waveguide

The *x-y* cross section of single slot waveguide is shown in Fig. 3.14, where two bricks of Si are placed symmetrically about *y*-axis with its length along *z*-axis creating one slot (s_v) in between and the whole SOI structure is filled with SiO_2 as a cladding material, composing this whole structure as commonly known, buried slot waveguide. Taking origin at the middle of slot, 'a_1' represents half of the slot width, ranging between $-a_1 < x < a_1$ or $|x| < a_1$ on *x*-axis. Similarly, $a_1 < |x| < b_1$ represents the brick on either side of middle slot.

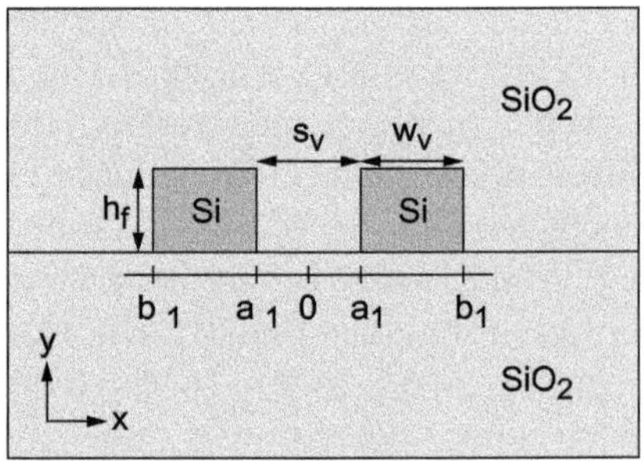

Fig. 3.14. Schematics of slot waveguide ($N = 1$).

The fundamental solutions of magnetic field (H_y) using TM mode equation are represented in Eq. 3.12.

$$H_y = \begin{cases} A\cosh(\gamma_s |x|) & |x| < a_1 \\ B_1\cos(\kappa_h |x - a_1|) + C_1\sin(\kappa_h |x - a_1|) & a_1 < |x| < b_1 \\ F\exp(-\gamma_c |x - b_1|) & |x| > b_1 \end{cases} \quad ...(3.12)$$

As per boundary conditions, the magnetic field (H_y) and its derivative must be continuous at the interface between two materials. Thus, the boundary conditions are applied at x = ± a and ± b as shown in Fig. 3.14, and the resulting variables B_1, C_1 and F are calculated as:

$$B_1 = A_1\cosh(\gamma_s a_1) \quad ...(3.13)$$

$$C_1 = A_1\left(\frac{\gamma_s}{\kappa_h}\right)\left(\frac{n_h}{n_s}\right)^2 \sinh(\gamma_s a_1) \quad ...(3.14)$$

$$F = B_1\cos(\kappa_h w_v) + C_1\sin(\kappa_h w_v) \quad ...(3.15)$$

The equation of electric field (E_x) using Eq. 2.20 and substituting the variables B_1, C_1 and F can be expressed as:

$$E_x = \begin{cases} \frac{A_1}{n_s^2} \text{Cosh}(\gamma_s |x|) & |x| < a_1 \\ \frac{1}{n_h^2} [B_1 \cos(\kappa_h |x - a_1|) + C_1 \sin(\kappa_h |x - a_1|)] & a_1 < |x| < b_1 \\ \frac{F}{n_c^2} \exp(-\gamma_c |x - b_1|) & |x| > b_1 \end{cases} \quad ...(3.16)$$

As discussed in earlier sections, one value of β is obtained from transcendental equation of single slot waveguide by equating and solving its expressions, which results in first order quasi-TE guided mode and its normalized mode profile is shown in Fig. 3.15. At this initial stage of deriving mode field equation, centrosymmetric geometry and parametric analysis of single slot waveguide can be further used to develop multiple slot waveguides.

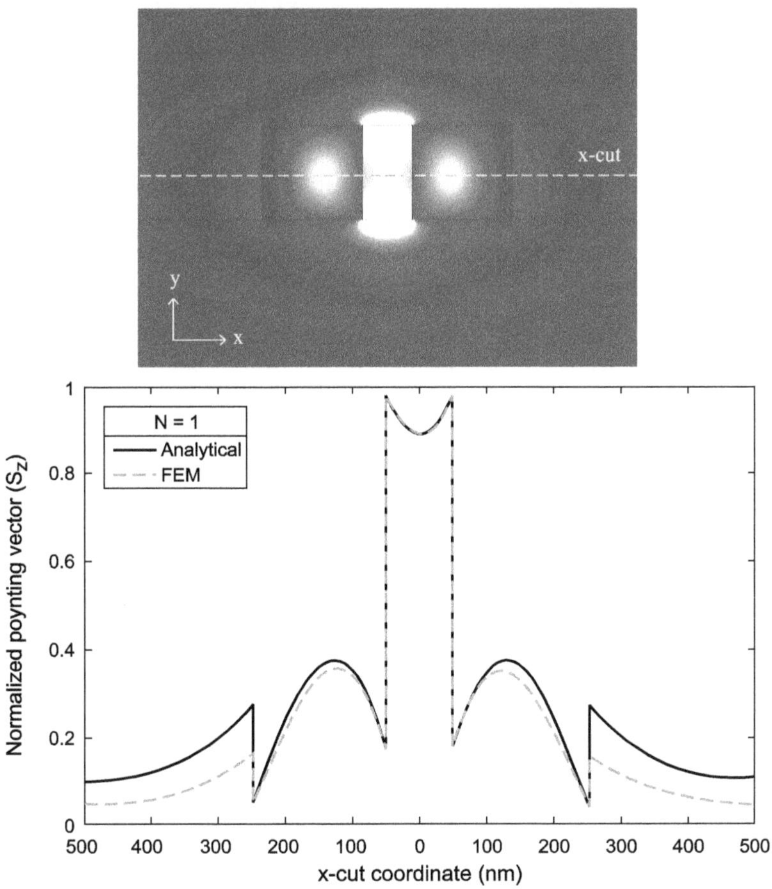

Fig. 3.15. Normalized mode profile of quasi-TE mode for $N = 1$: (a) *x-y* plane poynting vector profile; and (b) *x*-cut S_z comparison.

3.5.2 Mode in 'Three' Slot Waveguide

To create a base of multiple slot waveguide, it is important to know that whether mode will propagate with same footsteps as of parent vertical single slot waveguide, when number of slots is increased while keeping width constant. To exhort this possibility of guiding optical power inside multiple slot waveguide, significant theoretical study with contemplating analytical results are required. In case of multiple slot waveguide, N number of odd slots are considered having middle slot at the centre of waveguide. So, to start, three number of slots with equal pitch ratio of $s_v/w_v = 0.5$ is considered. The schematic of waveguide consisting of three slot and four brick is shown in Fig. 3.16.

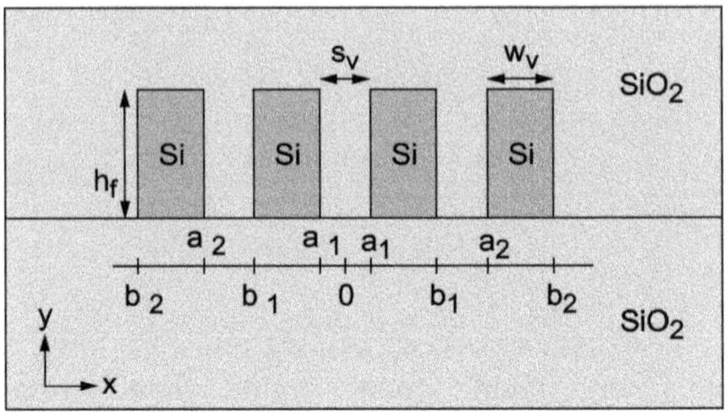

Fig. 3.16. Schematics of slot waveguide ($N = 3$).

Mode equations are derived by taking $S_v = 50$ nm and $W_v = 100$ nm. Thus, the overall width of waveguide structure consisting of 3 slots and 4 bricks is 550 nm, which is comparable to waveguide width of 500 nm in the previous case of single slot structure. The solution of magnetic field from TM mode given in Eq. 2.13 for each region is obtained as:

$$H_y = \begin{cases} A_1 \text{Cosh}(\gamma_s |x|) & |x| < a_1 \\ B_1 \cos(\kappa_h |x - a_1|) + C_1 \sin(\kappa_h |x - a_1|) & a_1 < |x| < b_1 \\ D_1 \text{Cosh}(\gamma_s \left|x - (\frac{a_2+b_1}{2})\right|) & b_1 < |x| < a_2 \\ B_2 \cos(\kappa_h |x - a_2|) + C_2 \sin(\kappa_h |x - a_2|) & a_2 < |x| < b_2 \\ F \exp(-\gamma_c |x - b_2|) & |x| > b_2 \end{cases} \quad ...(3.17)$$

Here, the values of B_1, C_1, D_1, B_2, C_2 and F are obtained by applying boundary condition at various slot brick interfaces, the results are represented in Eq. 3.18 to 3.23.

$$B_1 = A_1 \cosh(\gamma_s a_1) \quad ...(3.18)$$

$$C_1 = A_1 \left(\frac{\gamma_s}{\kappa_h}\right)\left(\frac{n_h}{n_s}\right)^2 \sinh(\gamma_s a_1) \quad ...(3.19)$$

$$D_1 = \frac{B_1 \cos(\kappa_h w_v) + C_1 \sin(\kappa_h w_v)}{\cosh\left(\gamma_s \left|b_1 - \left(\frac{a_2 + b_1}{2}\right)\right|\right)} \quad ...(3.20)$$

$$B_2 = D_1 \cosh\left(\gamma_s \left| a_2 - \left(\frac{a_2 + b_1}{2}\right)\right|\right) \quad ...(3.21)$$

$$C_2 = C_2 = D_1 \left(\frac{\gamma_s}{\kappa_h}\right)\left(\frac{n_h}{n_s}\right)^2 \sinh\left(\gamma_s \left| a_2 - \left(\frac{a_2 + b_1}{2}\right)\right|\right) \quad ...(3.22)$$

$$F = B_2 \cos(\kappa_h w_v) + C_2 \sin(\kappa_h w_v) \quad ...(3.23)$$

The corresponding value of electric field is derived and represented in Eq. 3.24. The theoretical mode profile along with FEM mode profile is represented and compared in Fig. 3.17. Through this figure, it is clearly observed that field intensity is not propagating inside silicon region. However, most of the field is found to be in the low refractive indexed region around silicon material. The evanescent field of individual silicon brick constructively interfere which increases its intensity inside slot region. However, huge variation in mode intensity can be seen inside slot itself with maximum being at the centre. This is due to discontinuity and disappearance of x component electric field inside silicon brick region, which result in splitting of oval shaped mode into its constituent as appeared in slot region. Also, since the constructive interference at the centre of structure is high with large amplitude of evanescent field, the intensity of the central slot is higher when compared to other slots.

$$E_x = \begin{cases} \frac{A_1}{n_s^2}\text{Cosh}(\gamma_s |x|) & |x| < a_1 \\ \frac{1}{n_h^2}\left[B_1 \cos(\kappa_h |x - a_1|) + C_1 \sin(\kappa_h |x - a_1|)\right] & a_1 < |x| < b_1 \\ \frac{D_1}{n_s^2}\text{Cosh}(\gamma_s \left|x - (\frac{a_2 + b_1}{2})\right|) & b_1 < |x| < a_2 \\ \frac{1}{n_h^2}\left[B_2 \cos(\kappa_h |x - a_2|) + C_2 \sin(\kappa_h |x - a_2|)\right] & a_2 < |x| < b_2 \\ \frac{F}{n_c^2}\exp(-\gamma_c |x - b_2|) & |x| > b_2 \end{cases} \quad ...(3.24)$$

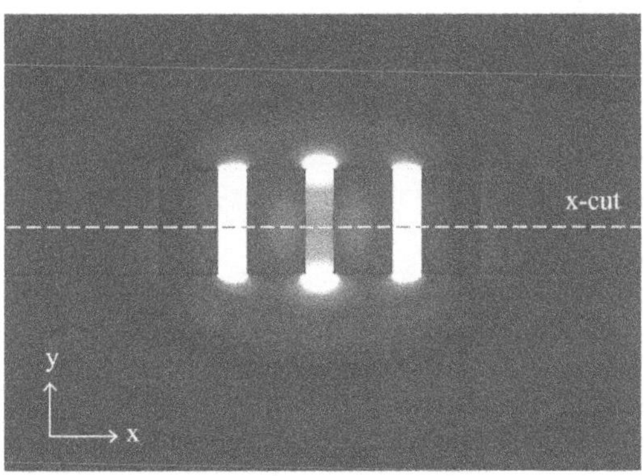

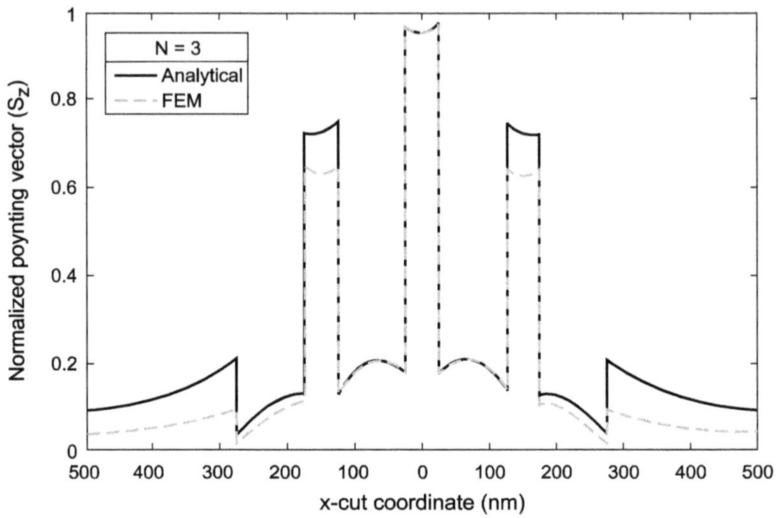

Fig. 3.17. Normalized mode profile of quasi-TE mode for $N = 3$: (a) *x-y* plane poynting vector profile; and (b) *x*-cut S_z comparison.

3.5.3 Mode in 'Five' Slot Waveguide

In the next stage of field profile derivation for '*N*' number of slots, where 1 and 3 number of slots is already derived earlier, will lead to the findings of more number of slots. Since, we are considering the schematics of multiple slot waveguide symmetric across origin in *x*-axis, with slot as the centre of waveguide. Thus, the value of N logically will be odd number. Up to this point field equations for $N = 1$ and 3 are acquired, now the mode field profile for $N = 5$ will be solved using similar way as of previous case. The schematic structure of waveguide containing of 5 slot regions with 6 brick regions is shown in Fig. 3.18.

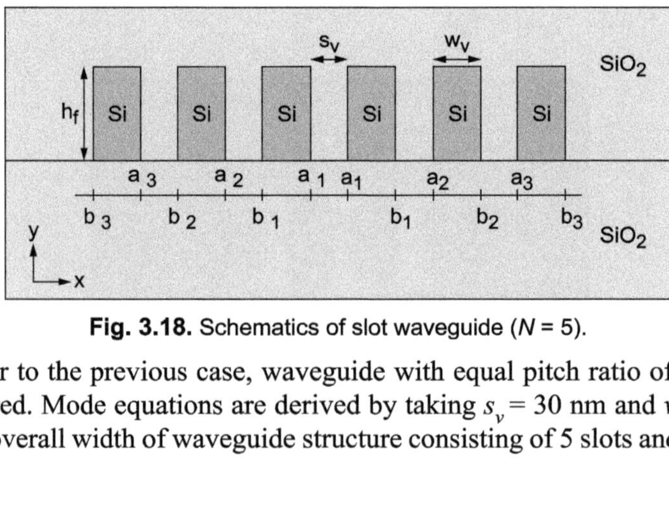

Fig. 3.18. Schematics of slot waveguide ($N = 5$).

Similar to the previous case, waveguide with equal pitch ratio of $s_v/w_v = 0.5$ is considered. Mode equations are derived by taking $s_v = 30$ nm and $w_v = 60$ nm. Thus, the overall width of waveguide structure consisting of 5 slots and 6 bricks is

510 nm, which is comparable to waveguide width of previous two cases. The x-cut coordinates of Fig. 3.18 is represented as:

$$a = \{a_1 \quad a_2 \quad a_3$$

$$a = \begin{cases} \dfrac{S_v}{2} & \dfrac{3S_v}{2} + w_v \quad \dfrac{5S_v}{2} + 2w_v \end{cases}$$

$$a_i = \left(\frac{2i-1}{2}\right)S_v + (i-1)w_v \qquad i = 1, 2, 3 \qquad \text{...(3.25)}$$

$$b = \{b_1 \quad b_2 \quad b_3$$

$$b = \begin{cases} \dfrac{S_v}{2} + w_v & \dfrac{3S_v}{3} + 2w_v \quad \dfrac{5S_v}{2} + 3w_v \end{cases}$$

$$b_i = \left(\frac{2i-1}{2}\right)S_v + (i)w_v \qquad i = 1, 2, 3 \qquad \text{...(3.26)}$$

The solution of magnetic field from TM mode as given in Eq. 2.13 for each region is obtained as:

$$H_y = \begin{cases} A_1 \operatorname{Cosh}(\gamma_s |x|) & |x| < a_1 \\ B_1 \cos(\kappa_h |x - a_1|) + C_1 \sin(\kappa_h |x - a_1|) & a_1 < |x| < b_1 \\ D_1 \operatorname{Cosh}(\gamma_s \left|x - (\frac{a_2+b_1}{2})\right|) & b_1 < |x| < a_2 \\ B_2 \cos(\kappa_h |x - a_2|) + C_2 \sin(\kappa_h |x - a_2|) & a_2 < |x| < b_2 \\ D_2 \operatorname{Cosh}(\gamma_s \left|x - (\frac{a_3+b_2}{2})\right|) & b_2 < |x| < a_3 \\ B_3 \cos(\kappa_h |x - a_3|) + C_3 \sin(\kappa_h |x - a_3|) & a_3 < |x| < b_3 \\ F \exp(-\gamma_c(|x - b_3|)) & |x| > b_3 \end{cases} \qquad \text{...(3.27)}$$

The magnetic field (H_y) and its derivative must be continuous at the interface between two materials. Thus, the boundary conditions are applied at $x = \pm a$ and $\pm b$ of each material boundaries, and the resulting solutions for variables are calculated as:

$$B_1 = A_1 \cosh(\gamma_s a_1) \qquad \text{...(3.28)}$$

$$C_1 = A_1 \left(\frac{\gamma_s}{\kappa_h}\right)\left(\frac{n_h}{n_s}\right)^2 \sinh(\gamma_s a_1) \qquad \text{...(3.29)}$$

$$D_1 = \frac{B_1 \cos(\kappa_h w_v) + C_1 \sin(\kappa_h w_v)}{\cosh\left(\gamma_s \left|b_1 - \left(\dfrac{a_2 + b_1}{2}\right)\right|\right)} \qquad \text{...(3.30)}$$

$$B_2 = D_1 \cosh\left(\gamma_s \left|a_2 - \left(\frac{a_2 + b_1}{2}\right)\right|\right) \qquad \text{...(3.31)}$$

$$C_2 = D_1\left(\frac{\gamma_s}{\kappa_h}\right)\left(\frac{n_h}{n_s}\right)^2 \sinh\left(\gamma_s\left|a_2-\left(\frac{a_2+b_1}{2}\right)\right|\right) \quad ...(3.32)$$

$$D_2 = \frac{B_2\cos(\kappa_h w_v)+C_2\sin(\kappa_h w_v)}{\cosh\left(\gamma_s\left|b_2-\left(\frac{a_3+b_2}{2}\right)\right|\right)} \quad ...(3.33)$$

$$B_3 = D_2\cosh\left(\gamma_s\left|a_3-\left(\frac{a_3+b_2}{2}\right)\right|\right) \quad ...(3.34)$$

$$C_3 = D_2\left(\frac{\gamma_s}{\kappa_h}\right)\left(\frac{n_h}{n_s}\right)^2 \sinh\left(\gamma_s\left|a_3-\left(\frac{a_3+b_2}{2}\right)\right|\right) \quad ...(3.35)$$

$$F = B_3\cos(\kappa_h w_v) + C_3\sin(\kappa_h w_v) \quad ...(3.36)$$

$$E_x = \begin{cases} \frac{A_1}{n_s^2}\mathrm{Cosh}(\gamma_s|x|) & |x|<a_1 \\ \frac{1}{n_h^2}\left[B_1\cos(\kappa_h|x-a_1|)+C_1\sin(\kappa_h|x-a_1|)\right] & a_1<|x|<b_1 \\ \frac{D_1}{n_s^2}\mathrm{Cosh}(\gamma_s\left|x-(\frac{a_2+b_1}{2})\right|) & b_1<|x|<a_2 \\ \frac{1}{n_h^2}\left[B_2\cos(\kappa_h|x-a_2|)+C_2\sin(\kappa_h|x-a_2|)\right] & a_2<|x|<b_2 \\ \frac{D_2}{n_s^2}\mathrm{Cosh}(\gamma_s\left|x-(\frac{a_3+b_2}{2})\right|) & b_2<|x|<a_3 \\ \frac{1}{n_h^2}\left[B_3\cos(\kappa_h|x-a_3|)+C_3\sin(\kappa_h|x-a_3|)\right] & a_3<|x|<b_3 \\ \frac{F}{n_c^2}\exp(-\gamma_c(|x-b_3|)) & |x|>b_3 \end{cases} \quad ...(3.37)$$

The electric field is derived and represented in Eq. 3.37. The theoretical mode profile along with FEM mode profile is represented and compared in Fig. 3.19. Again, this proves that the silicon bricks are still leaky, and the evanescent field of individual silicon brick constructively interfere and result to increase in intensity inside slot region. However, the mode intensity is not same in all the slots, the intensity of central slot region is maximum as compared to other slots. This is due to discontinuity and disappearance of x component electric field inside silicon brick region, which result in splitting of oval mode shape in to its constituent as appeared in slot region. Also, the constructive interference at the centre of structure is high with large amplitude of evanescent field, so the intensity of the central slot is higher when compare to other slots.

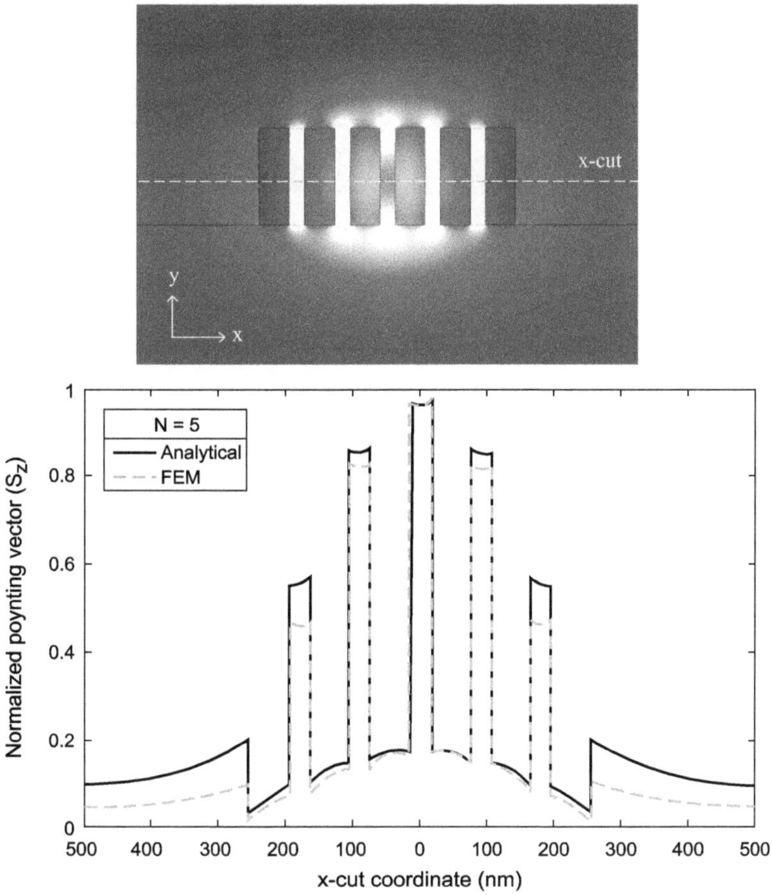

Fig. 3.19. Normalized mode profile of quasi-TE mode for $N = 5$: (a) *x-y* plane poynting vector profile; and (b) *x*-cut S_z comparison.

The multiple slot waveguide with three slots is commonly studied, although increased slot number is found to be having many more research applications, attracting various researchers for sensing and biomedical applications. Thus, this study will turn in to milestone if we would develop a universal equation for multiple slot waveguide and implement the same to SNORW waveguide.

Table 3.2. Data for vertical multiple slot waveguide

Number of SiO_2 slots	Size of slot	Number of Si bricks	Width of brick	Aspect ratio (s_v/w_v)	Waveguide total width
1	100 nm	2	200 nm	0.5	500 nm
3	50 nm	4	100 nm	0.5	550 nm
5	30 nm	6	60 nm	0.5	510 nm

❑❑❑

4 Theoretical Analysis of Silicon Nanowire Waveguide

This chapter is organized to gives a brief review of theoretical modal investigation of Silicon Nanowire Optical Rectangular Waveguide (SNORW). The mode equations are derived for SNORW using the results of multiple slot waveguide condition. In multiple slot vertical waveguide configuration, the centro symmetricity is present. The same theoretical analysis of multiple slot waveguide to N number of slots is extended for mode field calculation of Silicon Nanowire Optical Rectangular Waveguide.

4.1 SILICON NANOWIRE WAVEGUIDE NEED

In the past two decade, optical waveguides using Silicon-On-Insulator (SOI) technology are attracting many researchers of Photonic Integrated Circuits (PICs). Among various types of optical waveguides, the rectangular waveguide category is the typical PIC structure having wave guiding phenomenon inside the high Refractive Index (RI) region surrounded by low RI material, which is as similar to the optical fiber working phenomenon. In further studies, the guiding phenomenon inside low RI material surprised various researchers (Almeida *et al.*, 2004), as the guiding principle of optical waves is found to be the opposite of optical fibers. The vast application of smaller waveguide and design components exploration have given momentum to work in the area of low refractive index guided photonics (Kazanskiy, Butt and Khonina, 2021; Cam Hoang, 2022). Thereafter, a new category of waveguide called 'slot waveguide' had been evolved, which is now under intense research with its application over all the areas of PIC's (Iqbal *et al.*, 2020). Since then, many researchers have worked on the theoretical investigation of optical slot waveguide proposing wave solutions(Almeida *et al.*, 2004; Xu *et al.*, 2004; Priye, Malviya and Mickelson, 2018; Steglich, 2018), where the guiding phenomenon inside low RI slot region is proven with its diverse applications. This slot waveguide is classified into various category which mainly depends on its structural configuration. Orientation based classification is known as vertical/horizontal slot waveguide. The guidance of quasi-TE mode inside vertical slot waveguide and quasi-TM mode inside horizontal slot waveguide is also been investigated by various researchers (Almeida *et al.*, 2004; Xu *et al.*, 2004; Sun *et al.*, 2007; Singh, 2020b). It is proven that the guiding phenomenon is directly

DOI: 10.1201/9781003783589-4

dependent on the physical dimension of waveguide and its material compositions. With the moving span of time, the evolution of a very special type of waveguide based on the working property of slot waveguide evolved, which is silicon nanowire-based waveguide (Khorasaninejad and Saini, 2010, 2011; Singh, 2020a). The waveguide formation using vertically aligned silicon wires makes it more susceptible with high surface to volume ratio of guiding mode pattern (Singh, Malviya and Priye, 2016; Singh and Priye, 2018). This motivates to investigate wave solution of silicon nanowire-based structure, which is not appeared in any literature so far.

The diameter of silicon nanowires (SiNWs) for the waveguide formation ranges from 2 nm to 200 nm. The fined structured fabrication of nanowires should be in appropriate orientation with accurate dimension to harness desired photonic characteristics when used in the silicon nanowire-based devices. Hence, the production is very costly, which becomes challenging task for the scientific community. In general, 'bottom-up' and 'top-down' are two broadly categorized approaches for the fabrication of vertical SiNWs. Researchers have proposed several 'bottom-up' approaches such as the laser ablation technique (Yang *et al.*, 2004), Molecular Beam Epitaxy (MBE) technique (Fuhrmann *et al.*, 2005), thermal evaporation technique (Pan *et al.*, 2005), metal-organic vapor-phase epitaxy (Dick *et al.*, 2005) and Vapor–Liquid–Solid (VLS) method (Wagner and Ellis, 1964; Westwater, 1997; Latu-Romain *et al.*, 2008). The various methods to grow SiNWs in 'top-down' approaches are Reactive Ion Etching (RIE) technique (Fu *et al.*, 2009), Electro-less Wet Chemical Etching (EWCE) technique (Qiu *et al.*, 2006; Sohi and Kahrizi, 2017), Metal-Assisted Chemical Etching (MACE) (Han, Huang and Lee, 2014) and lithographic technique (Martínez, Martínez and Garcia, 2010). These silicon nanowire growth approaches require a clean chamber and are performed under very high temperature inside vacuum. However, these fabrication approaches are time consuming and expensive procedures to attain the most requisite criteria of having accurate orientation and precise dimension.

Nanowires based photonic devices are smaller in dimension with excellent operational performance (Fathi Aghdam, Liao and Huang, 2017; Yoon *et al.*, 2017; Hsin, Wu and Wang, 2019; Tong *et al.*, 2020). Silicon nanowires can be used in solar cell, due to its broadband absorption capacity along with very low reflectance (Peng *et al.*, 2005; Garnett and Yang, 2010). Nanowires on a large surface exhibit excellent light trapping property and demonstrate great potential in the use of absorber and photovoltaic applications (Venkatesan *et al.*, 2020). The surface of nanowire bed acts as anti-reflectance for light-intensity and a good option for carrier photon collections (Srivastava *et al.*, 2010). Silicon nanowire with its applications is explored in various fields of photonics, such as photovoltaic cells (Peng *et al.*, 2005), lithium batteries (Chan *et al.*, 2008), field effect transistors (Goldberger *et al.*, 2006), and photo-detectors (Bae *et al.*, 2010). Apart from applications in photonics, silicon nanowires are also applicable in various chemical (Cui, 2001), gas (Singh, Malviya and Priye, 2016), and biosensor devices (Abdul Rashid *et al.*, 2013; Singh and Priye, 2018). In addition, a closely packed bunch of silicon nanowires are preferred for guiding of optical waves, which has been demonstrated in this chapter with theoretical justification.

4.2 SNORW STRUCTURE

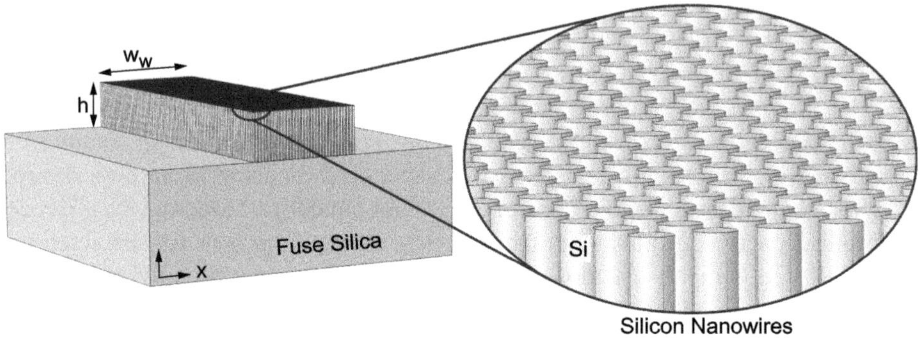

Fig. 4.1. 3D view of a Silicon Nanowire Optical Rectangular Waveguide (SNORW).

A Silicon nanowire-based waveguide structure is fabricated using two major steps. In the initial step, an array of silicon nanowire is grown over a layer of silicon oxide substrate by maintaining its desired physical parameters. After achieving the desired height of SiNWs, the final step is to remove the undesired portion of the SiNWs by the etching process. Consequently, the remaining bunch of vertical SiNWs work as a waveguide, which have a continuous constant aperture within all the silicon nanowires. The SiNWs having a vertical rod-shaped sub-wavelength structure, when arranged in an array pattern forms a waveguide known as 'Silicon Nanowire Optical Rectangular Waveguide' (SNORW). A 3D view of silicon nanowire optical rectangular waveguide without any cladding material is shown in Fig. 4.1. A SNORW consists of an array of high refractive indexed silicon wires/rods, which are arranged periodically along x and z direction above substrate. These rods are few nanometers in diameter, thus known as nanowires. The SNORW, when used for PIC based communication, the cavity between nanowires are filled with low refractive indexed material such as fused silica. The substrate material is also taken as fused silica, thus making the whole structural arrangement compatible with Silicon-On-Insulator (SOI) technology. The height of each silicon nanowire is 'h' and the effective waveguide width is 'w_w'.

$$w_w = (s_v \times N) + w_v \times (N + 1) \qquad ...(4.1)$$

Where, 'w_v' is the diameter of individual nanowire and 's_v' is the cavity spacing between two adjacent nanowires and 'N' is the number of cavity spacings along x-axis of SNORW cross-section. The combination of SiNWs and the cavity filled with fused silica, collectively make this structure as buried SNORW.

A Silicon Nanowire Optical Rectangular Waveguide (SNORW) consists of an array of high indexed silicon wires/rods arranged in a forest like structure, with numerous low refractive indexed narrow slits in between. This waveguide is a newly developed class of waveguides that has received significant attention by many researchers' worldwide, due to its promising applications. It is important to investigate the properties of these silicon nanowires and analyse its interaction with optical waves by deriving the condition at which guiding could be possible inside the SNORW structure.

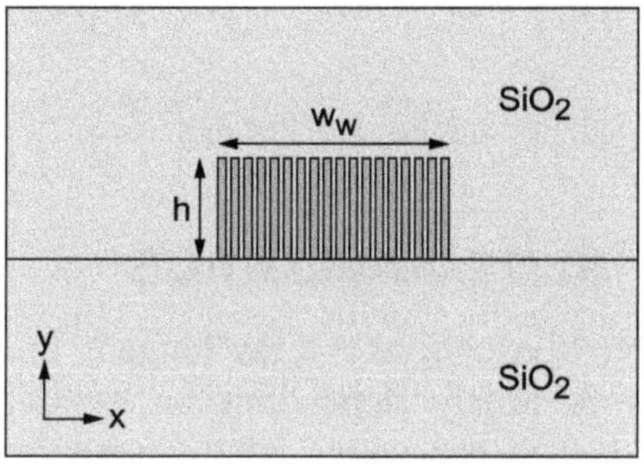

Fig. 4.2. Schematics of SNORW.

In SNORW, high refractive indexed nanowires are vertically aligned in an arrayed pattern on the surface of substrate material, SiO_2. The thin nanoslots between the nanowires are filled with SiO_2 material, which collectively make this structure as buried SNORW. Nanoslots along with upper cover layer is known as cladding which is been occupied by fused silica as shown in Fig. 4.2. This SNORW structure has 'N' (odd) number of nanoslots generated from periodically aligned 'N+ 1' number of silicon nanowires having dimensions of subwavelength. The diameter of each wire is taken as w_v, with the gapping of s_v and the height of h as shown in Fig. 4.3.

Fig. 4.3. SNORW for theoretic analysis.

The x-cut coordinates of SNORW with N nanoslots as shown in Fig. 3.18 is represented with its generalized coordinates as:

$$a = \left\{ a_1, \quad a_2, \quad a_3, \quad \ldots \quad a_{\frac{N+1}{2}} \right.$$

$$a = \left\{\frac{S_v}{2},\quad \frac{3S_v}{2}+w_v,\quad \frac{5S_v}{2}+2w_v,\quad \ldots\quad \frac{NS_v}{2}+\frac{(N+1)w_v}{2}\right.$$

$$a_i = \left(\frac{2i-1}{2}\right)S_v+(i-1)w_v \quad i=1,2,\ldots\left(\frac{N+1}{2}\right) \qquad \ldots(4.2)$$

$$b = \left\{b_1,\quad b_2,\quad b_3,\quad \ldots\quad b_{\frac{N+1}{2}}\right.$$

$$b = \left\{\frac{S_v}{2}+w_v,\quad \frac{3S_v}{2}+2w_v,\quad \frac{5S_v}{2}+3w_v,\quad \ldots\quad \frac{NS_v}{2}+\frac{(N+1)w_v}{2}\right.$$

$$b_i = \left(\frac{2i-1}{2}\right)S_v+(i)w_v \; i=1,2,\ldots\left(\frac{N+1}{2}\right) \qquad \ldots(4.3)$$

4.3 MODE IN 'N' SLOT WAVEGUIDE

The derivation of mode field equations is a challenge in this case as the interaction between field intensity is highly dependent on an overlapped mode region generated by the individual Si nanowires. The constructive interference evolved from the evanescent field of Si nanowires, inside nanogap region will result in low refractive index guiding waveguide. The resulting analytical solutions for the transverse magnetic (TM) field are derived region wise by using the solution of Eq. 2.13 as:

$$H_y = \begin{cases} A_1\,\mathrm{Cosh}(\gamma_s|x|) & |x|<a_1 \\ B_1\cos(\kappa_h|x-a_1|)+C_1\sin(\kappa_h|x-a_1|) & a_1<|x|<b_1 \\ D_i\,\mathrm{Cosh}\left(\gamma_s\left|x-\left(\frac{a_{i+1}+b_i}{2}\right)\right|\right) & b_i<|x|<a_{i+1} \\ & i=1,2,\ldots\left(\frac{N-1}{2}\right) \\ B_{i+1}\cos(\kappa_h|x-a_{i+1}|)+C_{i+1}\sin(\kappa_h|x-a_{i+1}|) & a_{i+1}<|x|<b_{i+1} \\ & i=1,2,\ldots\left(\frac{N-1}{2}\right) \\ F\exp\left(-\gamma_c\left|x-b_{\frac{i+1}{2}}\right|\right) & |x|>b_{\frac{i+1}{2}} \\ & i=N \end{cases}$$

...(4.4)

Where, $B_1 = A_1\cosh(\gamma_s a_1)$ ...(4.5)

$$C_1 = A_1\left(\frac{\gamma_s}{\kappa_h}\right)\left(\frac{n_h}{n_s}\right)^2\sinh(\gamma_s a_1) \qquad \ldots(4.6)$$

$$D_i = \frac{B_i \operatorname{Cos}(\kappa_h w_v) + C_i \operatorname{Sin}(\kappa_h w_v)}{\cosh(\gamma_s \left| b_i - (\frac{a_{i+1}+b_i}{2}) \right|)} \quad i = 1, 2, \ldots (\tfrac{N-1}{2}) \qquad \ldots(4.7)$$

$$B_{i+1} = D_i \cosh(\gamma_s \left| a_{i+1} - (\tfrac{a_{i+1}+b_i}{2}) \right|) \quad i = 1, 2, \ldots (\tfrac{N-1}{2}) \qquad \ldots(4.8)$$

$$C_{i+1} = D_i (\tfrac{\gamma_s}{\kappa_h})(\tfrac{n_h}{n_s})^2 \sinh(\gamma_s \left| a_{i+1} - (\tfrac{a_{i+1}+b_i}{2}) \right|) \quad i = 1, 2, \ldots (\tfrac{N-1}{2}) \qquad \ldots(4.9)$$

$$F = F = B_{\frac{i+1}{2}} \operatorname{Cos}(\kappa_h w_v) + C_{\frac{i+1}{2}} \operatorname{Sin}(\kappa_h w_v) \quad i = N \qquad \ldots(4.10)$$

$$E_x = \begin{cases} \dfrac{A_1}{n_s^2} \operatorname{Cosh}(\gamma_s |x|) & |x| < a_1 \\ \dfrac{1}{n_h^2} \left[B_1 \cos(\kappa_h |x - a_1|) + C_1 \sin(\kappa_h |x - a_1|) \right] & a_1 < |x| < b_1 \\ \dfrac{D_i}{n_s^2} \operatorname{Cosh}(\gamma_s \left| x - (\tfrac{a_{i+1}+b_i}{2}) \right|) & b_i < |x| < a_{i+1} \\ & i = 1, 2, \ldots (\tfrac{N-1}{2}) \\ \dfrac{1}{n_h^2} \left[B_{i+1} \cos(\kappa_h |x - a_{i+1}|) + C_{i+1} \sin(\kappa_h |x - a_{i+1}|) \right] & a_{i+1} < |x| < b_{i+1} \\ & i = 1, 2, \ldots (\tfrac{N-1}{2}) \\ \dfrac{F}{n_c^2} \exp(-\gamma_c \left| x - b_{\frac{i+1}{2}} \right|) & |x| > b_{\frac{i+1}{2}} \\ & i = N \end{cases} \qquad \ldots(4.11)$$

4.4 SNORW OPTIMIZATION

In this structure, the *x-y* cross-section of SNORW is considered for illustration of waveguide mode and the *x-z* top view cross-section is considered for illustration of wave guiding capability. The geometrical parameters of this SNORW structure are optimized for its confinement factor and feasibility. Confinement factor of waveguide with respect to the number of nanowires arranged in one array of *x-y* cross-section, the diameter of individual nanowire, and the cavity spacing between two adjacent nanowires are represented in Fig. 4.4(*a*), 4.4(*b*), and 4.4(*c*), respectively. '*N*' number of cavities with '*N*+1' number of nanowires are arranged in one array of x-y cross-section. The value of '*N*+1', 'w_v', and 's_v' are chosen as 34, 10 nm and 5 nm, respectively. These optimum values are the minimum possible waveguide geometrical parameters after which confinement factor will show a constant behavior. The ratio of 's_v' and 'w_v', known as pitch ratio, is calculated as 0.5. This arrayed pattern is repeating itself in z-direction with constant pitch ratio of 0.5. For illustrative purpose, optimized 34 nanowires and 33 nanogaps are considered in one array of SNORW along *x*-axis. Thus, the x-y cross-section consists N = 33 nanogaps and leads to effective waveguide width of 505 nm as per Eqn. 1. The effective waveguide width of ~0.5 μm will be comparable to various single mode supported waveguides, such as bulk SOI rectangular waveguides (Aalto *et al.*, 2004) and slot waveguides (Zengzhi *et al.*, 2015). This waveguide is designed mainly for silicon photonic integrated device-to-device communication on chip level. The whole SNORW structure is compatible with SOI technology. Therefore, fused silica (SiO_2) material is used in substrate as well as cladding portion and nanowires are made-up of silicon (Si) material.

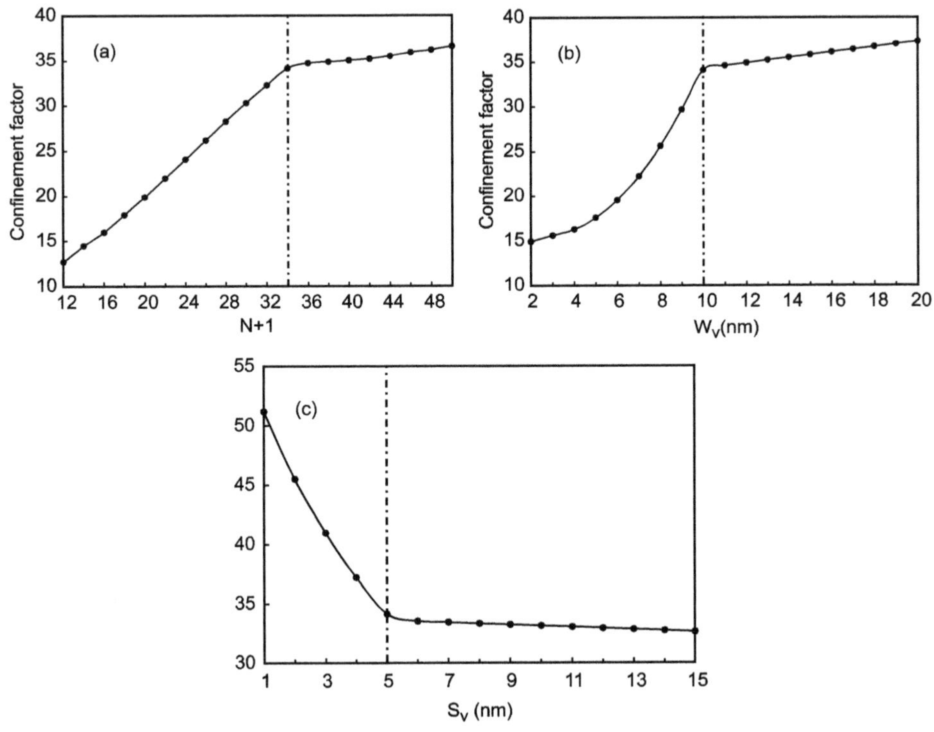

Fig. 4.4. Confinement factor with respect to (a) '*N*'+1, (b) 'W_v', and (c) 'S_v'; of SNORW.

Validation of mode type and filed distribution of the waveguide structure is crucial before evaluating its performance as a communicating integrated device. Thus, the *x-y* cross-section of SNORW is analyzed for its electric and magnetic field distribution pattern to judge its supported mode category. For this, the mode properties of this SNORW structure is studied and analyzed using full-vectorial finite element method. The wavelength chosen for this analysis is 1550 nm, which is mostly used wavelength in all types of optical communication scenarios (Zhou *et al.*, 2020). The refractive index used for SiO_2 and Si at 1550 nm wavelength is calculated through Sellmeier's equation as 1.4444 (Agrawal, 2010; Singh, 2020*a*) and 3.4752 (An *et al.*, 2014), respectively. The height of all silicon nanowire is chosen as 220 nm to meet the current standard height for SOI structures (Han *et al.*, 2020).

4.5 SNORW FIELD DISTRIBUTION

The computation of electric and magnetic fields guiding inside SNORW structure is simulated by taking minimum triangular mesh element size of 0.1 nm and maximum element size of 1 nm. The minimum mesh element size is $1/50^{th}$ the size of smallest SNORW parameter, that is s_v. Also, the maximum element size is $1/5^{th}$ the smallest waveguide parameter. Therefore, the precision and effectiveness of the achieved results is expected to be higher as compared to structure with large mesh size. The computed results for x, y and z components of electric and magnetic fields are represented in Fig. 4.5, Fig. 4.6 and Fig. 4.7, respectively.

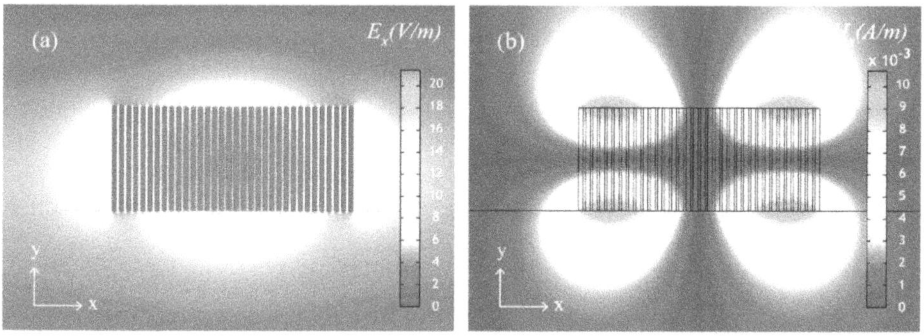

Fig. 4.5. X-component of (a) Electric field, and (b) magnetic field; profile of SNORW.

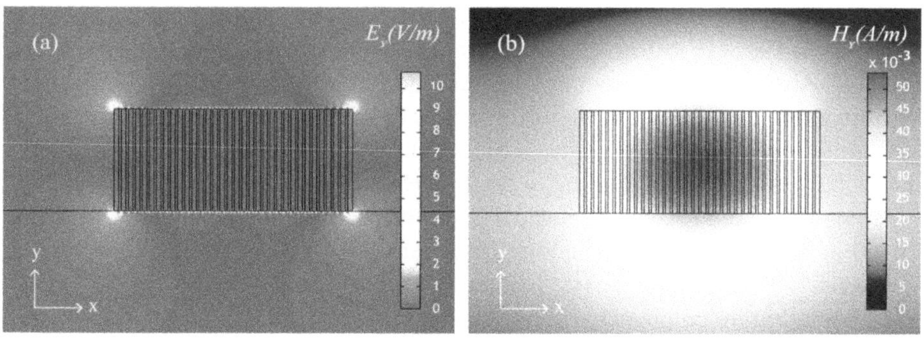

Fig. 4.6. Y-component of (a) Electric field, and (b) magnetic field; profile of SNORW.

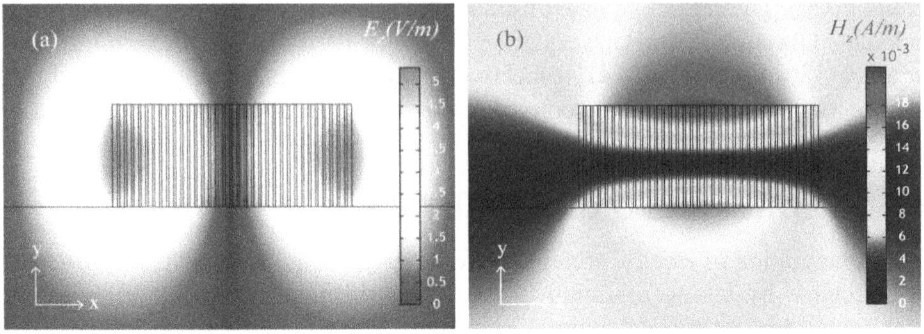

Fig. 4.7. Z-component of (a) Electric field, and (b) magnetic field; profile of SNORW.

The absence of electric or magnetic field in the direction of propagation shows the phenomena known as pure transverse, which means that the components are perpendicular to some common reference. If z-direction is considered as direction of wave propagation, then Transverse Electric (TE) field means absence of electric field in z-direction. For a pure TE mode, guided wave has only longitudinal magnetic field (i.e. $H_z \neq 0, E_z = 0$). Also, In TE mode H_x, E_y and H_z are non-vanishing, but E_x, H_y and E_z are vanishing because the electric field has only a transverse component. Therefore, TE-mode is not the guiding case in SNORW, since all three components of E as well as H are present, which is clear from Fig. 4.5, Fig. 4.6 and Fig. 4.7. On the other hand, Transverse Magnetic (TM) field represent the absence of magnetic field in z-direction. Pure TM field having guided waves with only longitudinal electric field (i.e. $E_z \neq 0$, $H_z = 0$). Also, in TM mode, the magnetic field has only a transverse component which result to non-vanishing E_x, H_y and E_z, but H_x, E_y and H_z are vanishing. Again, due to presence of all the field components, mode of SNORW is not similar to perfect TM mode. However, the intensity of fields such as E_x, H_y and E_z are much higher than that of H_x, E_y and H_z, thus, it can be inferred as partial TM type of mode. This is clear with the field intensity results for SNORW as shown in Fig. 4.5 to Fig. 4.7. Furthermore, transverse electromagnetic (TEM) mode is also a category, in which the z-component of both electric and magnetic field are absent (i.e. $H_z = 0, E_z = 0$). Generally, in TEM, mode field distribution, the waveguides have very small refractive index difference between core and cladding, and will often have weak guiding capability. However, in case of SNORW the two materials used are silicon and fused silicon, have refractive index of 3.4752 and 1.4444 at 1550 nm, which is having large refractive index difference of 2.0308. Moreover, the z-component of both E and H field is present. Therefore, TEM mode is not a guiding mode inside SNORW structure.

In dielectric material-based waveguide where no metallic boundary exists, the evolution of hybrid modes comes in to picture. These hybrid modes having presence of both electric and magnetic field components in z-direction (i.e. $H_z \neq 0$, $E_z \neq 0$). Hybrid modes are mainly categorized in to EH (quasi-TM) and HE (quasi-TE) modes (Orlando, 1992). Dominating mode component in EH and HE modes are H_z and E_z, respectively (Kapoor and Singh, 2000). In SNORW, the normal component of the electric field strength approximately vanishes at the boundary of SNORW as shown in Fig. 4.5. In addition, both electric and magnetic field components in z-direction are present as shown in Fig. 4.7, thus this waveguide is satisfying the conditions for existence of hybrid mode. The intensity of E_x and H_y is much higher than that of E_y and H_x respectively, which shows that this SNORW behaves like TM mode. However, with presence of all field component with dominancy of E_z over H_z field component this SNORW will work as HE (quasi-TE mode).

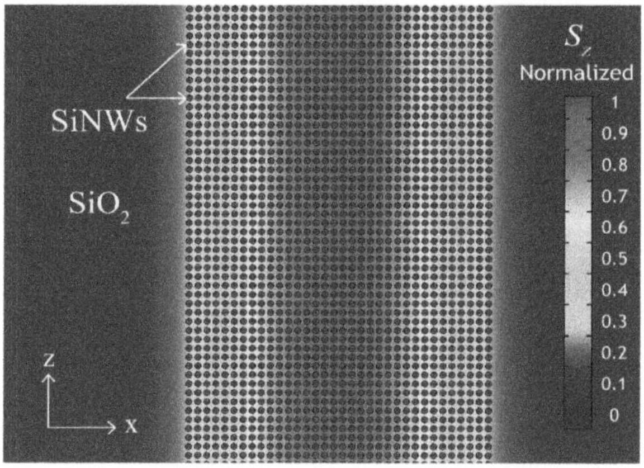

Fig. 4.8. x-z top view cross-section for wave intensity guiding in SNORW.

The hybrid HE (quasi-TE) mode is propagating inside this nanowire structure with dominancy of E_z over the other z-component of field, and presence of non-vanishing E_x, H_y. The resultant poynting vector after cross product of E_x and H_y will support optical intensity to propagate in z-direction. The homogeneous repetition of silicon nanowire pattern along z-axis is simulated and result for intensity guiding in SNORW structure is shown in Fig. 4.8. In SNORW, high optical power is confined inside the low RI nano-slots between silicon nanowires, which obey the same principle of wave guidance as of slot waveguide. However, the overall mode guiding pattern is quite similar to bulk rectangular waveguide. Therefore, it can be inferred from all the above simulated results that this SNORW structure has hybrid working phenomenon of both slot waveguide as well as rectangular waveguide.

4.6 SNORW MODE ANALYSIS

Like previous section of multiple slot waveguide, SNORW with equal pitch ratio of $s_v/w_v = 0.5$ is also considered in this section. Derived mode equations are illustrated by taking an example in which $S_v = 5$ nm and $W_v = 10$ nm. Thus, the overall width of waveguide structure consisting of 33 nanogaps and 34 wires is 505 nm, which is comparable to waveguide width of previous three cases of multiple slot waveguide. The modal properties of this SNORW structure are studied and analysed by full vectorial finite element method algorithm as shown in Fig. 4.9(a).

Because of the field discontinuity at high indexed contrast interfaces, the optical field is strongly enhanced in nanoslots, which is low indexed region of SNORW. The theoretical mode profile along with FEM mode profile is represented and compared in Fig. 4.9(b).

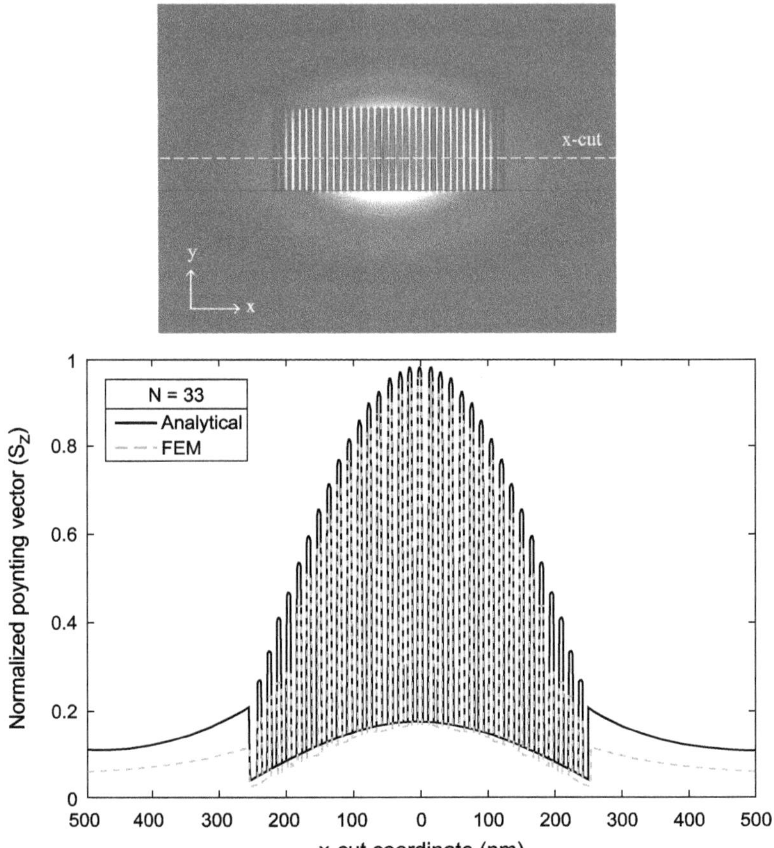

Fig. 4.9. Normalized mode profile of quasi-TE mode for $N = 33$: (a) *x-y* plane poynting vector profile; and (b) *x*-cut S_z comparison.

Again, these results prove that silicon wires in arrayed geometry can guide optical field. The evanescent field of individual silicon nanowire constructively interfere and increase the intensity inside nanoslot region. However, the mode intensity is discrete in all the nanogaps, with maximum being at the centre of waveguide. This discreteness in value is due to the discontinuity and disappearance of x component of electric field inside silicon nanowire. As a result, splitting occurs in oval mode shape as appeared in x-cut intensity plot, Fig. 4.9. Also, since the constructive interference at the centre of structure is high with large amplitude of evanescent field, the intensity of the central slot is higher when compared to other slots.

4.7 COMPARISON OF WAVEGUIDES

The results shows that the SNORW structure supports guiding of wave as quasi-TE mode. The performance of this waveguide at wavelengths other than 1550 nm need to be evaluated for sustainability of this structure as PIC component. Therefore, the effective refractive index (n_{eff}) and the mode propagation constant (β) of this SNORW is calculated from 1460 nm to 1625 nm wavelengths. The optical wavelength range from 1460 nm to 1625 nm are very important to utilize in the worldwide communication standards. The range of 1460-1625 nm wavelength is divided into three bands, i.e. S, C and L bands. The Short (S)-band starts from 1460 nm and ends at 1530 nm, commonly used in passive optical network systems as a downstream wavelength. The Conventional (C) - band starts from 1530 nm and ends at 1565 nm, mostly used in ultra-long-haul transmission systems with WDM and EDFA technologies. The Long (L) - band starts from 1565 nm and ends at 1625 nm, mainly used to expand the capacity of DWDM optical networks. These wavelength bands are suitable to perform almost all the operations of a PIC in photonic domain. The wavelength ranging from 1460 nm to 1625 nm with 1 nm resolution is used for calculation of n_{eff}, and β, as shown in Fig. 4.10(a). These solutions are then used for calculation of quasi-TE mode inside SNORW structure. The corresponding confinement factor (Γ_s) of field intensity inside this structure is plotted in Fig. 4.10(b). The confinement lies within range from 40.11% (at 1460 nm) to 30.98% (at 1625 nm), which shows linearly decreasing nature with increase in wavelength.

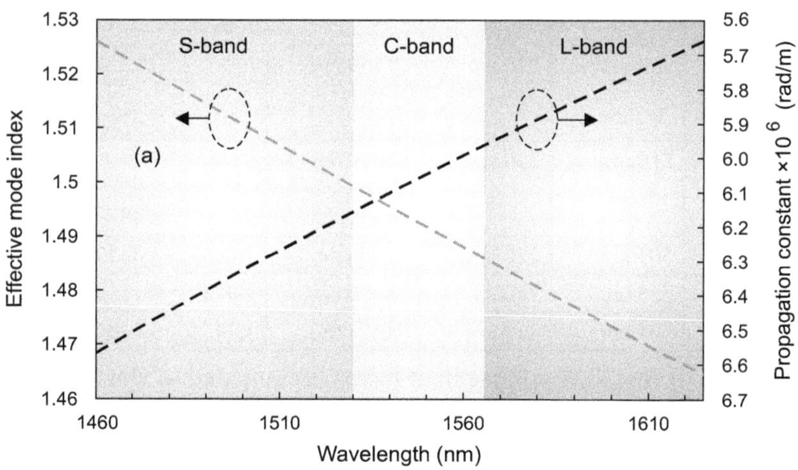

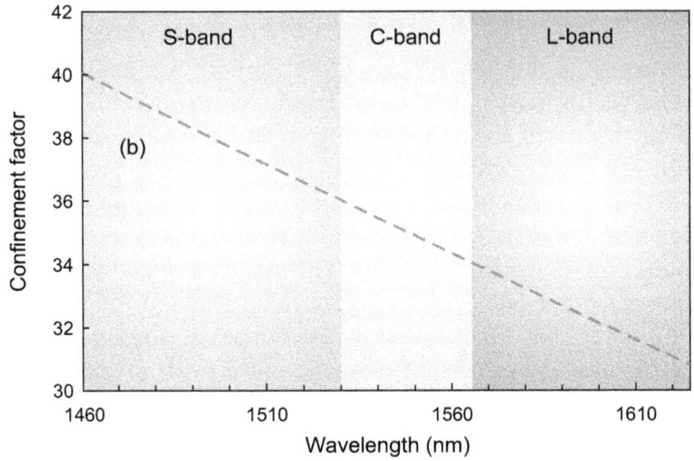

Fig. 4.10. (a) SNORW n_{eff} and β with wavelength; and (b) confinement of SNORW with wavelength.

Table 4.1. Confinement Factor Comparison of SOI Waveguides

Waveguide type	Effective waveguide width	Waveguide height	Operating wavelength (λ)	Low RI	High RI	Γs
Slot waveguide	420 nm	320 nm	1550 nm	1.53	3.476	~10 %
Slot waveguide	410 nm	300 nm	1550 nm	1.46	3.48	~31 % (Almeida *et al.*, 2004)
Slot waveguide	500 nm	220 nm	1550 nm	1.444	3.4752	25.02 % (Singh, 2020b)
Slot waveguide	516 nm	220 nm	1550 nm	1.70	3.48	21.6 % (Steglich, 2018)
SNORW	505 nm	220 nm	1460 nm	1.4451	3.4823	40.11 %
SNORW	505 nm	220 nm	1550 nm	1.444	3.4752	34.88 %
SNORW	505 nm	220 nm	1625 nm	1.4431	3.4701	30.98 %

The value of SNORW confinement factor is compared to slot waveguides in Table 4.1, which shows that this SNORW structure is supporting the propagation of mode at wavelengths of S, C and L-bands. The poynting vector of SNORW at wavelength of 1460 nm, 1550 nm and 1625 nm with *x*-cut coordinates are plotted in Fig. 4.11.

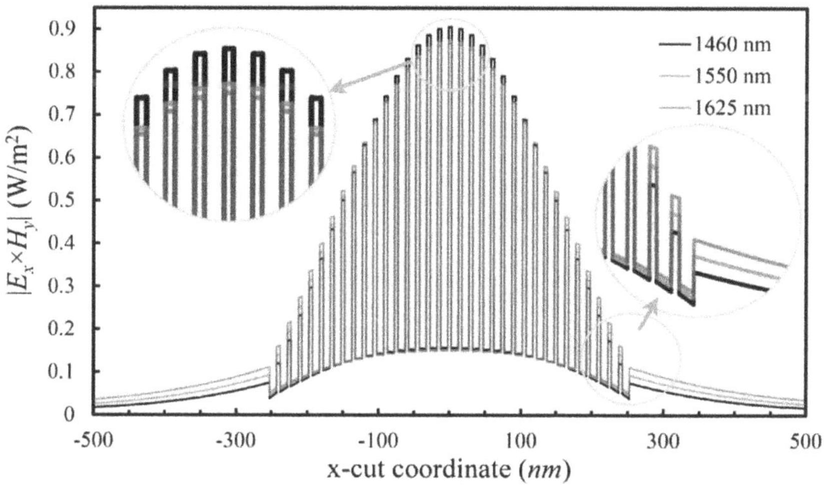

Fig. 4.11. SNORW Poynting vector profile for 1460nm, 1550nm and 1625nm.

The center of the waveguide structure have the maximum intensity. As the wavelength increases from 1460 nm to 1625 nm, the intensity decreases at the center region of SNORW structure. This shows that the lower wavelength is having high confinement in the waveguide region. In addition, lower wavelength is also having low evanescent field intensity outside the waveguide region. However, within the waveguide region, the evanescent field of individual silicon nanowire constructively interfere and increase the intensity inside nanoslot region. The mode intensity is discrete in all the nanogaps, with maximum being at the center of waveguide. The discreteness in mode intensity value is due to the discontinuity and disappearance of x component of electric field inside silicon nanowire. As a result, splitting occurs in oval-shaped field distribution profile. Also, the constructive interference at the center of structure is high as a result of high intensity at the central nanoslot as compared to other nanoslots. These results prove that the SiNWs in arrayed geometry can guide optical field intensity with excellent operational performance for SOI based PICs.

❑❑❑

5 Dispersion Analysis of SNORW

In this chapter, a special type of subwavelength waveguide structure which was introduced in earlier chapter based on vertical patterned silicon nanowires named as Silicon Nanowire Optical Rectangular Waveguide (SNORW) is considered for dispersion engineering. The equivalent refractive index of the vertical wired array deposited on top of a fused silica can be adjusted to have a wide range of mode propagation constants values for the fundamental mode. As a consequence, the rate of mode transmission to the subwavelength waveguide structure can be finely controlled to obtain flattened dispersion profile. Optical properties including group velocity dispersion coefficient and effective index of SNORW is tailored by varying waveguide parameters at wavelengths of S, C and L international telecommunication union bands. Waveguide dispersion mainly depends upon the change in effective refractive index with different geometry of the waveguide and operating wavelengths (Zhang, Yue, Xiao-Li, *et al.*, 2010; Gunasundari *et al.*, 2014). Therefore, modal effective index and dispersion are systematically studied with key parameters of the SNORW structures. Dispersion and mode loss are also analyzed with the real and imaginary part of effective refractive index. The theoretical results show that SNORW have different and unique dispersion properties which could influence their potential applications.

5.1 NEED OF DISPERSION ENGINEERING

Recently, subwavelength periodic structures having period smaller than the wavelength of light have attracted large number of researchers from the field of integrated photonic circuits. Since a subwavelength structure can be replaced with a homogeneous material without any reflection or diffraction (Halir *et al.*, 2015), it has been utilized to improve the performance of various photonic devices, such as directional couplers (Halir *et al.*, 2012), multimode interference couplers (Maese-Novo *et al.*, 2013), polarization splitters and rotators (Pérez-Galacho *et al.*, 2013; Guan *et al.*, 2014; Xiong *et al.*, 2014, 2015), adiabatic couplers (Yun *et al.*, 2016), and filtering devices (Wang, Glesk and Chen, 2014). Other applications that might benefit from subwavelength structures are those that require chromatic dispersion engineering. Chromatic dispersion is critically important in both linear and nonlinear applications (Agrawal, 2007; Yue *et al.*, 2010). Despite the fact

DOI: 10.1201/9781003783589-5

that integrated dispersive elements with flat and high dispersion are desired for applications such as tunable delay lines (Sharping *et al.*, 2005), analog to digital converters (Bhushan, Coppinger and Jalali, 1998), optical correlators (Geraghty *et al.*, 2008), and dispersion compensators (Zhang *et al.*, 2009), they have been explored in just few works.

Vertically aligned silicon nanowires on silicon-dioxide substrate functions as an optical waveguide, due to its ability to confine high electric field within lower indexed nanoslots between silicon nanowires (Khorasaninejad and Saini, 2010, 2011). Various devices that are fabricated on SOI platform such as optical switches (Fischer, ZInke and Petermann, 1995), ring resonators (Baehr-Jones *et al.*, 2004), optical amplifiers (Jones *et al.*, 2005; Foster *et al.*, 2006), modulators (Reed *et al.*, 2010) and interferometers (Zhao *et al.*, 1995) can be made with silicon nanowire waveguide as well which are topics of future research. Optical field intensity and confinement factor are found to be quite high inside nanoslots of SNORW as compared to slot optical waveguide. Thus, introducing SNORW geometries for comprehensive study of both flat and low dispersion is absolutely invaluable. Therefore, rigorous effort is required for optimization in the design of SNORW to improve its communication performance through proper mode field distribution. Waveguide dispersion plays a vital role in optical communication (Zheng, Iqbal and Liu, 2008; Iqbal, 2014; Ashok, Lee and Shin, 2016; Mialdun and Shevtsova, 2017).

In general, there are a few different aspects in terms of dispersion engineering that one may desire to tailor based on an intended application: (i) the sign of dispersion (i.e., normal or anomalous); (ii) the number of zero-dispersion wavelengths (ZDWs) and their locations; (iii) the slope of dispersion; (iv) the flatness of dispersion profile, which is defined as the total dispersion variation divided by bandwidth; and (v) the flattened dispersion bandwidth. However, simultaneous control of all these aspects is quite challenging. Nonetheless, these properties can be tailored through adjusting and optimizing the structural parameters of waveguides (Zhang, Yue, Beausoleil, *et al.*, 2010; Zhang *et al.*, 2012; Guo *et al.*, 2016; Jafari *et al.*, 2016). Particularly, to control the flattened dispersion bandwidth, two different approaches have been utilized (Jafari *et al.*, 2016): (i) increasing the waveguide cross-section size at the expense of higher total dispersion variation and lower flatness, or (ii) considering a cladding material with a higher RI from the allowable range of low-indexed material in multilayer or bi-layer waveguides (i.e., lower index contrast between silicon and the low-index material).

Ultrafast transmission can be achieved by lowering the dispersion effect for a particular wavelength. Bit rate can be improved with lesser inter-symbol interference in case of lower dispersion effect. Besides low dispersion, the flat dispersion effect is also an important assessment to be considered in WDM & DWDM networks. Fast-changing and strong waveguide dispersion in silicon waveguide create difficulties in achieving flat dispersion. But with the evolution of low RI guided waveguide, this difficulty is removed (Zhang *et al.*, 2015). In general, flat dispersion profile for broad wavelength range can be obtained either by increasing the waveguide cross-sectional size or by replacing the waveguide cladding material with high RI from the range of low-indexed material (Zhang, Yue, Beausoleil, *et al.*, 2010; Zhu *et al.*, 2012).

At signal processing end, inter-symbol interference should be as low as possible. For which, dispersion coefficient needs to be flat and low. Flat dispersion profile will lead to equal broadening of DWDM wavelengths, creating the overlapping between adjacent signals in same proportion. In fact, flat and low dispersion in any optical waveguide is desirable to achieve in international telecommunication union band applications, such as signal processing, femtosecond pulse or octave spanning spectral broadening (Zhang, Yue, Beausoleil, *et al.*, 2010; Zhang *et al.*, 2012, 2015; Zhu *et al.*, 2012). In SOI waveguide, phase matching (Zhu *et al.*, 2012) and supercontinuum generation (Zhang *et al.*, 2012) are efficient for achieving flat and low dispersion profile. Also, flat and low dispersion in waveguide have potential to explore new nonlinear properties like parametric amplification, wavelength conversion, ultrafast pulse manipulation, ultra-broadband signal processing, modulation and propagation (Tien *et al.*, 2010; Zhang, Yue, Xiao-Li, *et al.*, 2010; Liu *et al.*, 2011). However, so far, characteristics of Group Velocity Dispersion (GVD) and dispersion parameter (D) with their relation to structural parameters of SNORW have not been explored till date.

5.2 UNDERSTANDING DISPERSION PHENOMENA

Dispersion represents a broad class of phenomena related to the fact that velocity of electromagnetic waves depends on its wavelength. Dispersion is used to describe the processes, which cause the deterioration of signals carried by the electromagnetic waves while propagating in an optical waveguide. This deterioration occurs because the different orders of guided mode pattern having different phase constants propagate with different velocities inside a guiding channel. Mainly, dispersion is divided into two categories, one is mode dispersion, and another is chromatic dispersion.

Temporal broadening of pulse at the end of multimode waveguide, due to the different path lengths traversed by different modes, is termed as mode dispersion. Since the rays which belong to different modes, will travel different distance along varying paths in a given point of time resulting in different arrival time of each ray. Modal dispersion can therefore be described as path length dispersion. Modal dispersion is the biggest cause of dispersion in multi-mode waveguide, but not in single mode waveguide. However, modal dispersion in SNORW waveguide is not occurring as the dimensions of waveguide is considered in such a way that allows only one mode to travel. Thus, mode dispersion is not encountered in our analysis due to the absence of more than one mode in SNORW. Since, mode dispersion analysis is now not requisite to the investigation on SNROW, intensive study of chromatic dispersion will be carried out in further sections.

Chromatic dispersion is the spreading of pulse because refractive index of a waveguide is a function of wavelength. The chromatic dispersion depends on characteristic feature of guiding material and is impossible to eliminate completely. However, it can be reduced to a certain level which could be used to operate SNORW efficiently. Material dispersion and waveguide dispersion are two types of chromatic dispersion. Material dispersion occur due to the presence of finite linewidth in the light source and due to the dependence of refractive index of the material on wavelength. On the other hand, waveguide dispersion is only important in single mode waveguide. It is caused by the fact that some light intensity travels inside the waveguide cladding as compared to most of the light intensity which travels inside the waveguide core. Various tweaks in the design of single mode waveguide can be used to overcome waveguide dispersion, and manufacturers are constantly refining their processes to reduce its effects. Waveguide dispersion describes the dependence of the effective refraction index (n_{eff}) on the propagating wavelength through the waveguide. The Group Velocity Dispersion (GVD) (β_2) rely on the wavelength and wavelength dependent effective mode index (n_{eff}) (Ghatak and Thyagarajan, 1989) that can be expressed as:

$$\beta_2 = \frac{\partial^2 \beta}{\partial \omega^2} = \frac{\lambda^3}{2\pi c^2}\frac{\partial^2 n_{eff}}{\partial \lambda^2} \quad ...(5.1)$$

where β is the propagation constant, ω is the angular frequency, n_{eff} is the effective mode index, λ is the wavelength and c is the speed of light in vacuum.

The waveguide which shows positive β_2, implies that the effective refractive index of waveguide increases with decrease in wavelength. Most commonly, this phenomenon is known as normal dispersion ($\beta_2 > 0$). But when the effective refractive index of waveguide increases with increase in wavelength, then β_2 will be negative and this phenomenon is termed as anomalous dispersion ($\beta_2 < 0$). In an identical way, materials showing positive β_2, indicate that the wave components at longer wavelengths travel faster than components at shorter wavelengths. However, shorter wavelengths will travel faster if the material is showing negative β_2. In addition, if the value of β_2 is larger, the broadening of optical pulse in a given time domain will be larger. It is said that the pulse is modulated positively (positively chirped), when the longer waves travel in a medium faster than the short ones (Maine *et al.*, 1988).

In an optical dielectric waveguide, wave propagation is limited by the boundary of optical channel in which wave is generally confined. Also, the wave propagation is well dependent on the refractive index of guiding material. These limitations to the waveguide parameters with their dependency on guiding material will be reflected from the effective refractive index (n_{eff}) of the waveguide. The mode propagation constant (β) of a waveguide is related to the wave vector of electromagnetic radiation (k_0) and the effective refractive index (n_{eff}) of waveguide as given in Eq. 5.2. Additionally, by expressing k_0 in terms of wavelength as shown in Eq. 5.2, it is confirmed that β will depend inversely on wavelength and directly on wavelength dependent effective index (Ghatak and Thyagarajan, 1989; Dattner and Yadid-Pecht, 2011).

$$\beta = n_{eff} k_0 = \frac{n_{eff}\omega}{c} = \frac{2\pi n_{eff}}{\lambda} \qquad ...(5.2)$$

The nature of dispersion can be defined by its dependency on the value of β_2. Broadening of optical pulse along the path of propagation can be found when β_2 is positive ($\beta_2>0$), which is described as normal dispersion. However, with negative value of β_2 ($\beta_2<0$), optical pulse propagates in a nonlinear waveguide without dispersion after assuming the form of optical soliton in anomalous dispersion case(Agrawal, 2007; Serna *et al.*, 2016). Standard dispersion parameter (D) depends on group velocity (V_g) (Ghatak and Thyagarajan, 1989) as:

$$D = \frac{\partial}{\partial\lambda}\left(\frac{1}{V_g}\right) = \frac{-2\pi c\beta_2}{\lambda^2} \qquad ...(5.3)$$

The value of D for a homogeneous medium covers both waveguide dispersion as well as material dispersion. Phase velocity (v_p) defines the velocity of the wave of constant phase for a given mode as:

$$v_p = \frac{c}{n} = \frac{\omega}{k} \quad \text{for a planar wave} \qquad ...(5.4)$$

$$v_p = \frac{\omega}{\beta} \quad \text{for a wave propagating in a waveguide} \qquad ...(5.5)$$

However, group velocity (v_g) depends mainly on the variation in group effective index (n_g) and it can be expressed in the form of wavelength and effective mode index as:

$$v_g = \frac{\partial\omega}{\partial\beta} = \left(\frac{\partial\beta}{\partial\omega}\right)^{-1} = \frac{c}{n_g} \qquad ...(5.6)$$

$$n_g = n_{eff} + \lambda\left(\frac{\partial n_{eff}}{\partial\lambda}\right) \qquad ...(5.7)$$

The dispersion parameter (Marchetti *et al.*, 2017) is determined in the form of second derivative of the wavelength dependent effective mode index.

$$D = \frac{-2\pi}{\lambda^2}\left(\frac{2\partial n_{eff}}{\partial\omega} + \frac{\omega\partial^2 n_{eff}}{\partial\omega^2}\right) = \frac{-\lambda}{c}\frac{\partial^2 n_{eff}}{\partial\lambda^2} \qquad ...(5.8)$$

Instead of β_2, the group velocity dispersion is often described as the dispersion coefficient (D) having unit of *ps/km-nm*. The connotation of 1 *ps/km-nm* is the temporal broadening of pulse in picoseconds after passage from 1km of optical channel, if the spectral width is of 1 nm.

5.3 MATERIAL DISPERSION

SOI guiding structures considered here have silicon acting as high refractive index material and fused silica acting as insulating substrate material. To proceed further with the waveguide structure, it is necessary to first investigate the effect of material and its dispersion for SOI based PIC structure. Wavelength dependent refractive index of silicon (n_{Si}) (An *et al.*, 2014) and fused silica (n_{SiO2}) (An *et al.*, 2014) are described by the Sellmeier's equation as:

$$n_{Si} = \sqrt{A_1 + \left(\frac{A_2}{\lambda^2}\right) + A_3\left(\frac{A_4^2\lambda^2}{\lambda^2 - A_4^2}\right)} \quad ...(5.9)$$

$$n_{SiO_2} = \sqrt{1 + \left(\frac{B_1\lambda^2}{\lambda^2 - B_2^2}\right) + \left(\frac{B_3\lambda^2}{\lambda^2 - B_4^2}\right) + \left(\frac{B_5\lambda^2}{\lambda^2 - B_6^2}\right)} \quad ...(5.10)$$

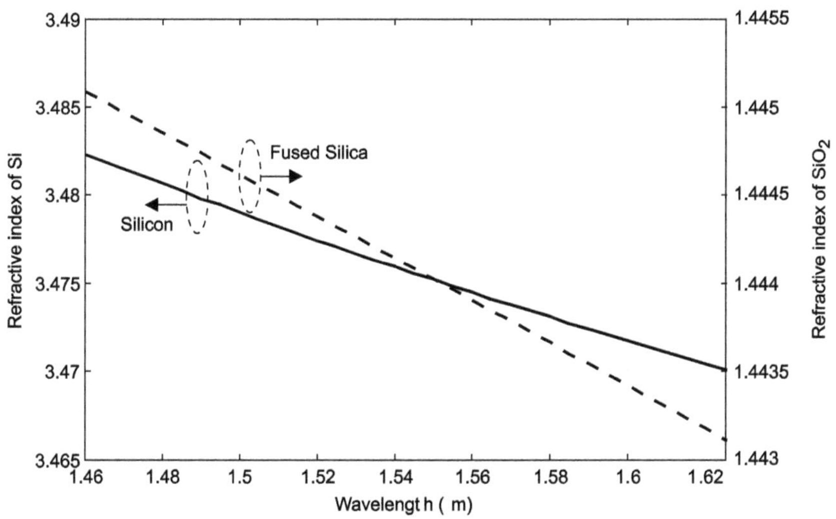

Fig. 5.1. Refractive index of silicon and fused silica vs wavelength.

Table 5.1. Data for Sellmeier's coefficients.

Sellmeier's coefficients for SiO_2	Values	Sellmeier's coefficients for Si	Values
B1	0.6961663	A1	11.6858
B2	0.0684043 μm	A2	9.39816×10^{-13} m^2
B3	0.4079426	A3	8.10461×10^{-3} m^{-2}
B4	0.1162414 μm	A4	1.1071×10^{-6} m
B5	0.8974794		
B6	9.8961610 μm		

The refractive index of silicon and fused silica at S, C and L-band wavelength is obtained from Eq. 5.9 and Eq. 5.10, respectively. The calculated refractive index of silicon and fused silica with the wavelength range between 1460 nm to 1625 nm are plotted in Fig. 5.1. From the Fig. 5.1, it is observed that the refractive index of both silicon and fused silica decreases with increase in wavelength.

Wavelength and wavelength dependent refractive index of material are used in the calculation of material dispersion parameter from Eq. 5.8, by replacing the modal effective index (n_{eff}) by material refractive index. The calculated material dispersion parameter of silicon and fused silica within wavelength range of 1460 nm to 1625 nm are plotted in Fig. 5.2. In this figure it can be observed that the material dispersion of both silicon and fused silica increases with wavelength. Moreover, the value of material dispersion for fused silica (SiO_2) is positive (>0) and material dispersion for silicon (Si) is negative (<0).

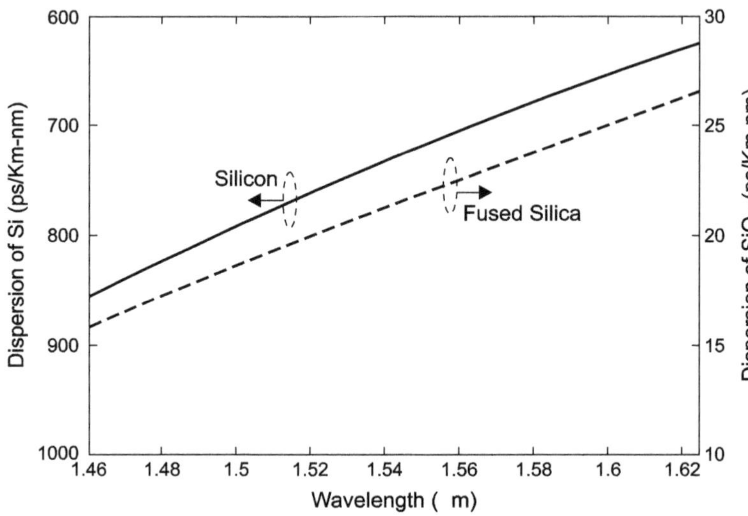

Fig. 5.2. Material dispersion of silicon and fused silica.

5.4 WAVEGUIDE DESIGN ANALYSIS

High refractive indexed nanowire rods are vertically aligned on the surface of substrate material in an arrayed pattern. The nanoslots between the nanowire rods are filled with a low refractive indexed material. Nanoslots along with upper cover layer is known as cladding and is occupied by air or fused silica. As shown in Fig. 5.3(a) Silicon-On-Insulator (SOI) technology based waveguides are considered in this study, where substrate and nanowires are made of fused silica and silicon respectively. The structure has periodic alignment of nanowires with subwavelength dimension and is named as Silicon nanowire optical rectangular waveguide (SNORW). The height of nanowire is h, the width of nanowire is d and the width of the nanoslot is g as shown in Fig. 5.3(b). The modal properties of the SNORW structure are studied and analysed by using full-vectorial finite element method algorithm. Analysis of modal propagation constant for principal guiding mode can be used to obtain modal effective index n_{eff}.

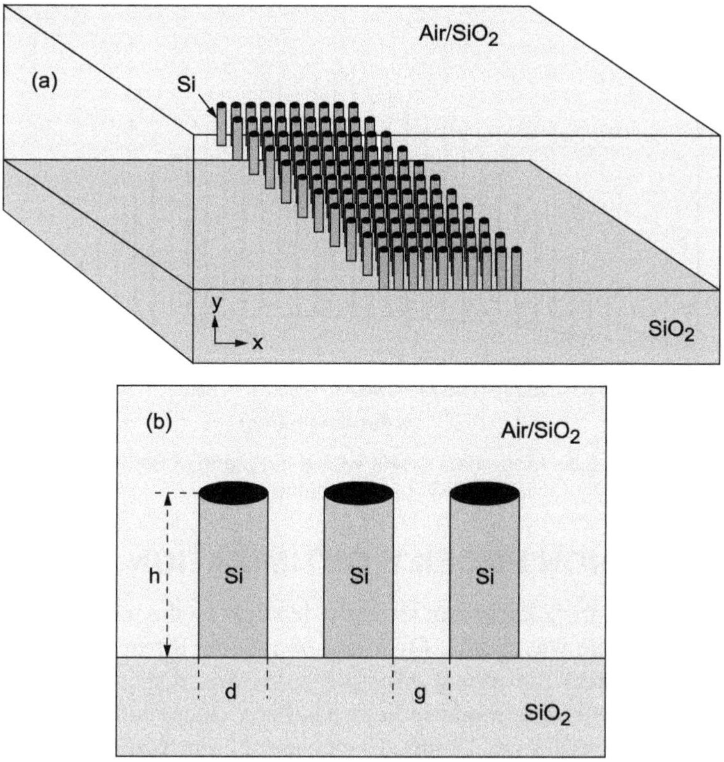

Fig. 5.3. Structure of SNORW. (a) 3D view; (b) Front view of three encircled nanowires.

In SNORW, quasi-TE polarization mode is observed to be the dominant optical propagation mode. Calculation of effective mode index for dominant mode is performed by full-vectorial finite element method, in which a non-uniform meshing of triangular elements by quadratic Lagrange function is used. Maximum mesh size is selected to be 50 nm while minimum mesh size is taken to be 0.1 nm, which is 1/20 of the lowest sized element. For illustration, mode field distribution profile along X-axis for $k = 33$, $g = 2$ nm, $d = 20$ nm and $h = 220$ nm at communication wavelength of 1550 nm is shown in Fig. 5.4. This figure shows that the confinement of field intensity is found inside the nanoslots of SNORW. Moreover, the pattern of intensity which is maximum at center and gradually decreases to both the waveguide ends will refer to as fundamental single mode SNORW. However, due to low refractive index of air as compared to SiO_2, air filled cladding have high normalized field intensity inside nanoslots than SiO_2 filled cladding. This study can be further extended to investigate the dispersion tailoring under the variation of relevant parameters of SNORW.

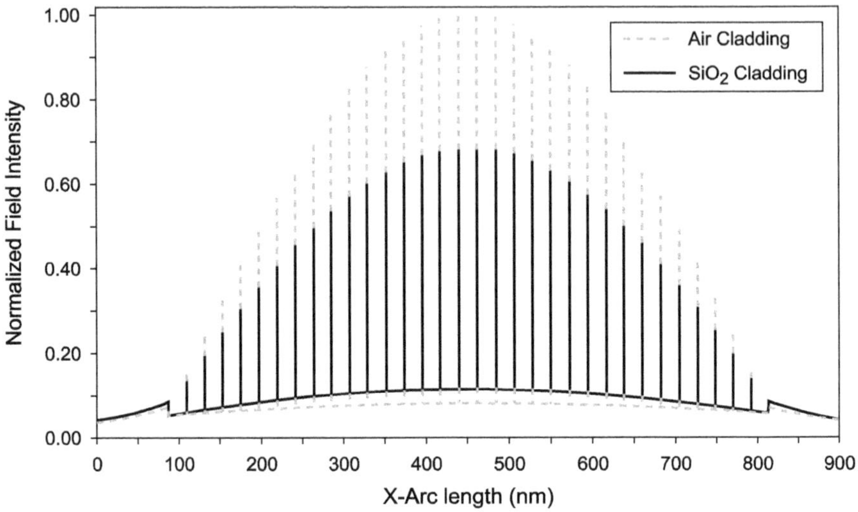

Fig. 5.4. Mode field profile with X-cut plane of SNORW for air/SiO_2 filled cladding region.

5.5 DISPERSION PROFILE OPTIMIZATION

In waveguide structures, dispersion strongly depends on the modal characteristics and parameters of the waveguide. Over and above, the diameter of nanorods *'d'* can significantly affect the modal effective index and dispersion parameter. To elucidate this, the effective mode indices has been calculated for SNORW with three different values of *d* i.e. 18 nm, 20 nm, and 22 nm while keeping the other SNORW design parameters constant. Wavelength dependent effective modal index for considered values of *d* are plotted in Fig. 5.5(a). As the diameter of silicon nanowire increases, the modal effective index of SNORW slightly shifts to higher values for an entire wavelength range of 1.46 μm to 1.625 μm. It can be seen from Fig. 5.5(a) that SNORW with fused silica-filled cladding has higher modal effective index than the air-filled cladding. Corresponding dispersions are calculated from the modal effective index values and the results are plotted in the Fig. 5.5(b). In this figure, the fused silica-filled waveguide has higher dispersion values than the air-filled waveguide. Hence, it is inferred that the magnitude of effective index and dispersion is significantly affected by the diameter of nanowire in waveguide.

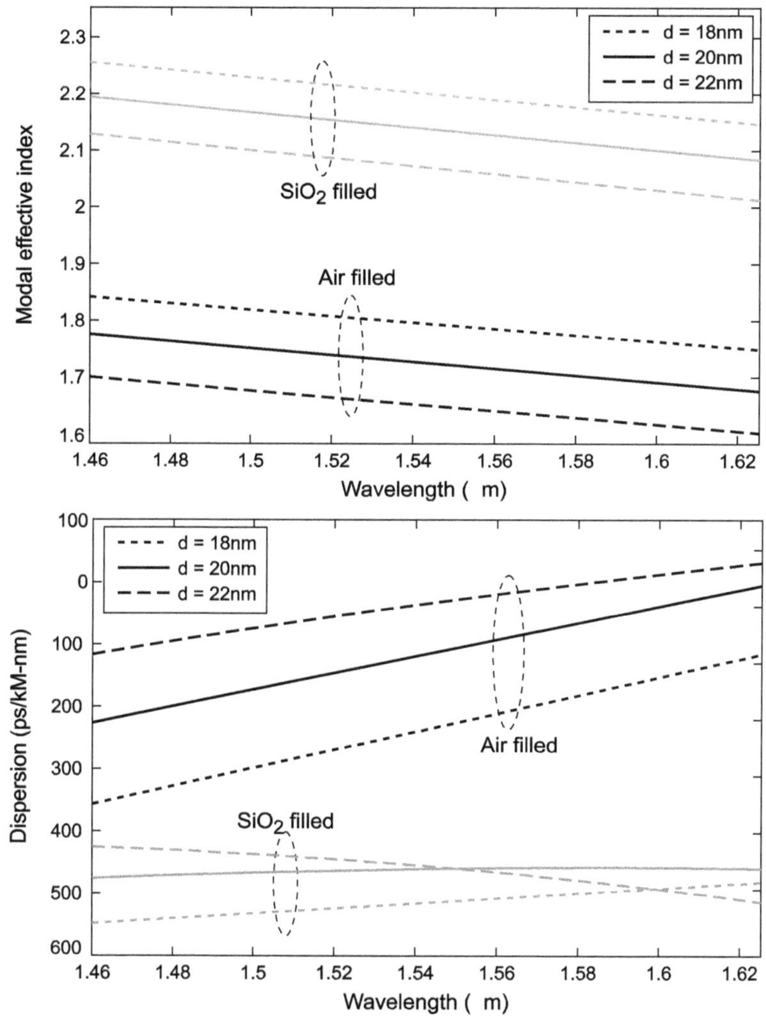

Fig. 5.5. (a) Modal effective index; (b) Dispersion; dependence on '*d*' with air & SiO_2 filled cladding.

Next, the effect of the gapping '*g*' between silicon nanowires and cladding material on the dispersion parameter is studied, where the diameter '*d*' of silicon nanowire rods is 20 nm while maintaining other dimensions of SNORW. To illustrate the effect, the modal effective index and dispersion are plotted against wavelength for three different values of '*g*' = 2 nm, 3 nm and 4 nm as shown in Fig. 5.6. The filling material of cladding portion is still considered as fused silica or air. As shown in Fig. 5.6(a), when *g* is increased, the modal effective index decreases. Air-filled cladding has less effective index than the fused silica-filled cladding. Corresponding dispersion values are shown in Fig. 5.6(b). It is seen from this figure that the dispersion values for air-filled cladding is high as compared to fused silica-filled cladding, whose slope is increasing as the value of *g* is increased. But for SiO_2 filled SNORW cladding, the dispersion curve is almost flat with lower dispersion value for smaller *g* size. For example, for SiO_2 filled cladding with *g* = 2 nm, the dispersion curve is almost flat with 455 ps/km-nm dispersion in the entire wavelength range of 1460 nm to 1625 nm. From the figure and discussion, it is clear that the value of effective index and dispersion is significantly affected by cladding material and gapping between silicon nanowires.

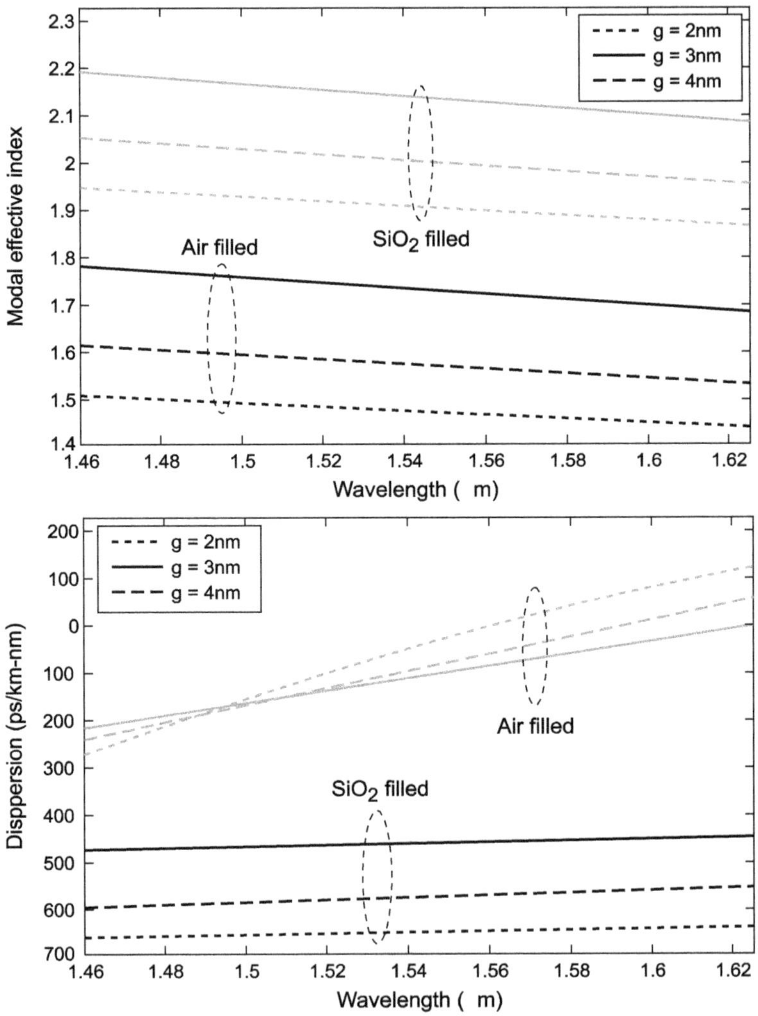

Fig. 5.6. (a) Modal effective index; (b) Dispersion; dependence on '*g*' with air & SiO_2 filled cladding.

Any change in number of nanowires '*k*' in one array of SNORW, can also affect the effective modal index and dispersion values. To verify this, variation of effective index and dispersion for three different values of k = 30, 33 and 35 is plotted in Fig. 5.7(a) and 5.7(b). The diameter of nanowire (d) and gapping between nanowires (g) are kept at 20 nm and 2 nm, respectively. As the value of 'k' increases along with refractive index of cladding, the modal effective index increases, due to the availability of larger portion of the field in nanoslots region. Corresponding dispersion curve is shown in Fig. 5.7(b). Higher dispersion value is obtained for higher refractive index of cladding. As the value of 'k' increases, the value of dispersion also increases.

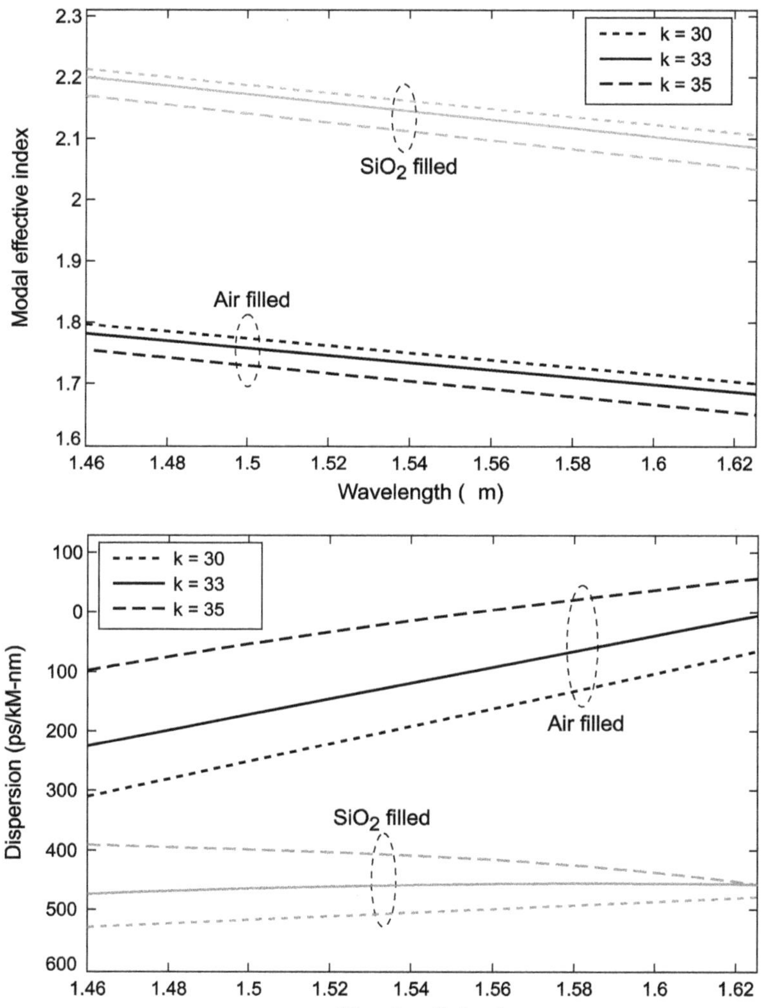

Fig. 5.7. (a) Modal effective index; (b) Dispersion; dependence on 'k' with air and SiO_2 filled cladding.

5.6 WAVEGUIDE DISPERSION

Now, based on the above methodology, the effective index and dispersion of SNORW is first calculated for the optimized waveguide dimension. The waveguide structure with height of nanowire (h) as 220 nm, diameter (d) of each nanowire as 20 nm, gapping (g) between nanowire as 2 nm and the number of nanowire (k) in one array as 33 are assigned. For analysis, the material of cladding is assumed to be air with refractive index of 1 or fused silica with refractive index obtained from Eq. 5.10. The modal effective index corresponding to the dominant mode is calculated under different wavelength conditions. In the calculation of modal effective index (n_{eff}), the refractive index of material is considered by using Sellmeier's Eq. 5.9 and Eq. 5.10 for silicon (Si) and fused silica (SiO_2), respectively. Calculated mode effective index is plotted against the wavelength in Fig. 5.8. From this figure,

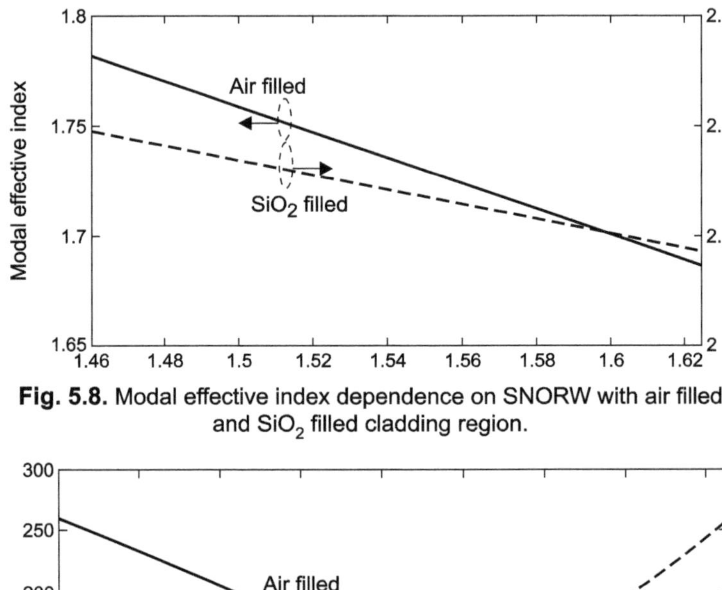

Fig. 5.8. Modal effective index dependence on SNORW with air filled and SiO_2 filled cladding region.

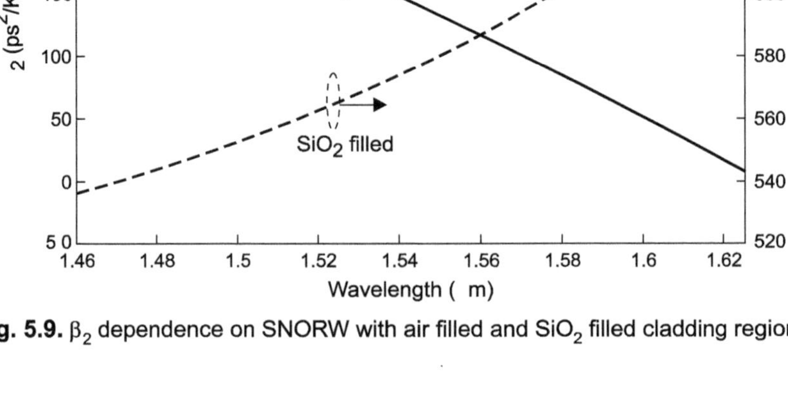

Fig. 5.9. β_2 dependence on SNORW with air filled and SiO_2 filled cladding region.

it can be observed that, SNORW with SiO_2 filled cladding has higher modal effective index than air-filled cladding, due to the higher refractive index of SiO_2 as compared to air. The corresponding Group Velocity Dispersion (GVD) (β_2) in ps^2/km is plotted in Fig. 5.9.

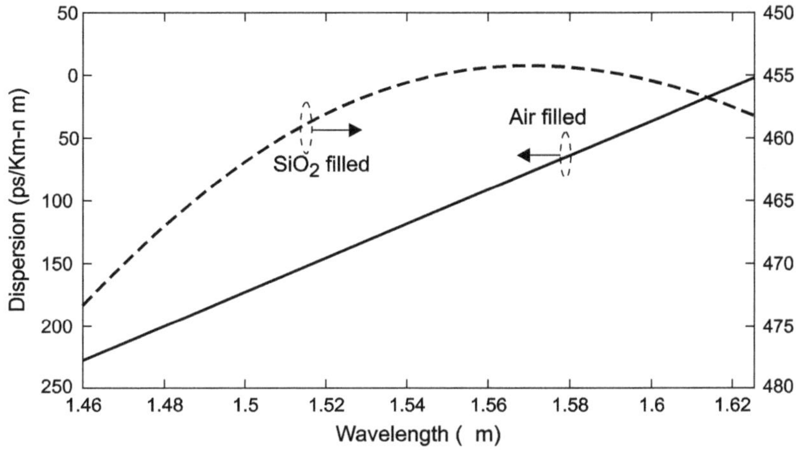

Fig. 5.10. Dispersion dependence on SNORW with air filled and SiO_2 filled cladding region.

The value of β_2 from Eq. 5.1 is positive for both air-filled as well as SiO_2 filled cladding. As stated in the theory section of this chapter, these findings suggest that the wave components will travel faster at longer wavelengths than at shorter wavelengths, for both air as well as SiO_2 cladding. However, the value of β_2 is higher for SiO_2 filled cladding than for air-filled cladding, for an entire wavelength range of 1.46μm to 1.625μm. This result indicates that the effective refractive index of waveguide with air-filled cladding decreases much faster with increase in wavelength as compared to SiO_2 filled cladding, which can be seen in Fig. 5.8.

The plot of dispersion parameter (*D*) in ps/km-nm of SiO_2 filled SNORW and air-filled SNORW modes against wavelength are shown in Fig. 5.10. In this figure, high dispersion value can be seen for SiO_2 filled cladding for entire S, C and L bands of wavelengths. In rectangular waveguide dispersion, field propagates in high refractive index, hence pulse broadening limits rate with which data transfer takes place. However, in slot waveguide and SNORW, field propagates in low refractive index slot and nanoslots, respectively.

5.7 WAVEGUIDE HEIGHT DEPENDENCY

Standard silicon height for Silicon-On-Insulator (SOI) fabrication are 220 nm and 330 nm. These values of heights are taken in consideration for better compatibility with other interconnecting optical waveguides. With SiO_2 as cladding material, modal effective index and dispersion profile is plotted for $h = 220$ nm and 330 nm in Fig. 5.11. Modal effective index of $h = 330$ nm is more than that of $h = 220$ nm, due to increase in percentage of composition of Si material having high refractive index than SiO_2. The dispersion parameter calculated by Eq. 5.8 is higher for $h = 330$ nm as compared to $h = 220$ nm. By observing Fig. 5.11, it can be inferred that n_{eff} and D increase with increase in height of silicon nanowires.

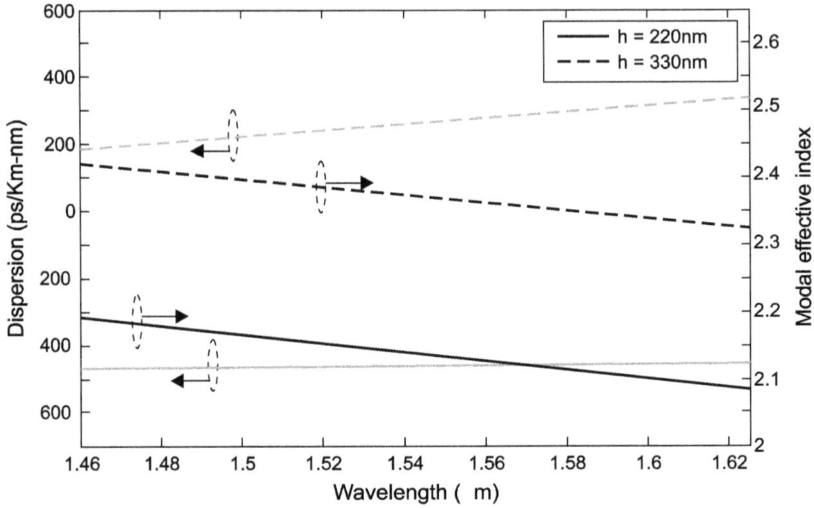

Fig. 5.11. Dispersion and modal effective index dependence on *h* with SiO_2 filled cladding.

5.8 MODE LOSS ESTIMATION

Besides dispersion, loss calculation is also important for optical waveguide. Mode pattern of SNORW is characterized by its field pattern and propagation constant. Effective refractive index depends upon propagation constant and free space wave number. Complex n_{eff} is obtained mainly due to losses. Mode loss is calculated from the imaginary part ($I_m(n_{eff})$) of the effective refractive index as (Khandokar *et al.*, 2016)

$$\text{Loss(dB/m)} = \frac{20}{\ln(10)} \frac{2\pi}{\lambda} \mathrm{I_m}(n_{eff}) \qquad ...(5.11)$$

Mode losses obtained from Eq. 5.11 for $h = 220$ nm and 330 nm of SNORW are plotted in Fig. 5.12. Losses are found low for the wavelength range of 1.49 μm to 1.57 μm. The focus of interest is on recently developed SOI technology of 220 nm.

From the above silicon nanowire optical rectangular waveguide (SNORW) results, it is clear that due the dependency of dispersion on waveguide, the structural configuration gives the flexibility to enhance the performance of SNORW.

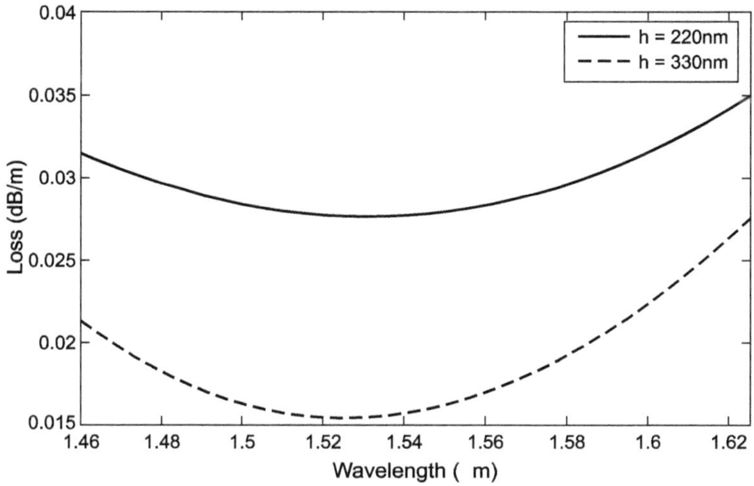

Fig. 5.12. Mode loss versus wavelength for SNORW with SiO_2 filled cladding region.

5.9 SUMMARY OF SNORW DISPERSION ANALYSIS

The utilization of SNORW as a subwavelength periodic structure based on SOI technology for dispersion engineering has been discussed. Effect of wavelength variation on modal effective index and dispersion are analyzed by taking two varieties of cladding material. In SNORW, light matter interaction occurs in an infinitesimal region with significant confinement factor value for varied dispersions. Results are obtained for a wide range of wavelengths which covers complete S, C and L optical bands. Most possible dispersion flatness is achieved from the highest refractive indexed material chosen from the allowable range of low indexed material. From results, it is found that dispersion behavior is almost flat for SiO_2 filled cladding. Effective index for the dominant mode of SNORW is been calculated by using Finite Element Method (FEM), and the corresponding dispersion is being calculated numerically. Mode field profile is analyzed for air-filled and fused silica-filled cladding SNORW. Considering high RI cladding as fused silica, dispersion and losses are evaluated for the standard silicon height of 220 nm and 330 nm as per SOI technologies. Decreasing *k*, *d*, and increasing *g* value, will result to decrease in field intensity of SNORW mode profile. A flattened dispersion over 1460 nm to1625 nm wavelength and a normal negative dispersion with low magnitude can be potentially useful for super-continuum generation, phase matching, wavelength conversion, parametric amplification, ultrafast pulse manipulation, high data rate telecommunication and enhancing nonlinear interactions of optical waves. This analysis on SNORW is a contribution to the future research focusing on compact nano-optical devices.

❑❑❑

6 Methane Sensing using SNORW Waveguide

In this chapter, a silicon nanowire based optical waveguide for sensing application is demonstrated with analysis. SOI based PIC structures using refractive index sensing is done for various waveguide structures. This chapter provides a detailed analysis of silicon photonic waveguide for methane gas sensing. In this chapter, a SNORW on 220 nm SOI platform is discussed for optical sensing of gases. Firstly, electric field intensity pattern of dominant quasi-TE mode is obtained for SNORW and compared with reported conventional slot and ridge slot waveguides (Zengzhi *et al.*, 2015). CF and waveguide sensitivity for conventional slot, ridge slot and SNORW are calculated and compared. To realize multiple SNORW on SOI platform, the compactness of these three waveguides are considered by comparing their physical dimensions for optimal performances. Lastly, its application in gas sensing is discussed with a typical example of methane gas which is considered to be hazardous in coalmines.

6.1 NEED OF METHANE SENSOR

In India, coal is major source of energy which is heavily utilized for various electrical power stations and industrial operations. Coal mining are of two types, one is overcast mining and another is underground mining. To get good quality of coal belt, underground mining is mainly preferred. Most of the coal mines in India is having various hazardous issues including roof collapse, rock burst, gas outburst and fire explosion. In these hazards, fire explosions and gas outburst are the most dangerous ones (Jiasheng Ni *et al.*, 2008), which mainly occurs due to methane trapping under the rock belts. Due to these hazards, the mining becomes more accident prone task as the mining depth increases every year (Liu *et al.*, 2013). In the underground coal mines, continuous monitoring of methane gas is strictly required due to highly explosive nature of methane and air mixture. The lower explosive limit of methane concentration is 5% and upper explosive limit is 15% as reported by DGMS (Directorate General of Mines Safety) India. Currently, most of the underground mines in India are using catalytic bed type methane sensing system which is usually capable of detecting 1-5% of methane level (Kumar, 2010). In order to protect people working under collieries and to avoid detonation inside mines, It is necessary to design a sensing system that

DOI: 10.1201/9781003783589-6

can predict the methane outburst by measuring exact concentration of methane fraction inside underground mines.

6.2 REFRACTIVE INDEX SENSING

During last decade, silicon waveguides are being extensively studied for different optical device applications (Law, Goldberger and Yang, 2004; Tasdemir *et al.*, 2015). Moreover, silicon waveguides also realized its potential application as optical sensors. Recently, silicon waveguide based optical sensors such as surface plasmon sensors (Nemova and Kashyap, 2007), double-slot-waveguide based ring resonator sensors (Dell'Olio and Passaro, 2007), Mach–Zehnder interferometer based sensors (Li and Tong, 2008), optical fiber sensors (Guo and Tong, 2008) and cavity ring-down structure based sensors (Humayun *et al.*, 2016) have been realized on SOI platform. Recently, References (Yan, Gargas and Yang, 2009; Khorasaninejad and Saini, 2010) reported that a bunch of properly designed nanowires can operate as an optical waveguide. To fabricate nanowire waveguides, bottom-up and top-down approaches are available (Chen *et al.*, 2013). However, bottom-up approach is preferred to fabricate devices of nanoscale dimensions (Chen *et al.*, 2013). In bottom up approach, silicon nanowire waveguide is fabricated by mounting an array of silicon nanowires on silicon oxide substrate and then undesired portion of the silicon wires are etched (Khorasaninejad and Saini, 2010). In such a silicon nanowire waveguide, high optical power is confined in the low indexed nanoslots present between silicon nanowires, which follows the same principle of guidance as slot waveguide (Almeida *et al.*, 2004).

Slot waveguide with air as cladding can be configured into an optical gas sensor. In Reference (Leonardis, 2012), simulation results for methane gas of concentration as low as 10% by volume in air was demonstrated using slot waveguide based ring resonator operating at 3.39μm. It works on the principle that RI of slot changes with exposure to gases resulting in variation of guiding parameters of the waveguide. For a waveguide to function as an efficient sensor, confinement factor (Xu *et al.*, 2004; Dell'Olio and Passaro, 2007) and sensitivity (Kargar and Chao, 2011; Zengzhi *et al.*, 2015) with respect to change in refractive index of slot should be high. Nanowire waveguide shows high CF due to high surface to volume ratio. It is also found to be ultrasensitive, highly selective with low power consumption and exceedingly responsive (Chen *et al.*, 2013).

Thus, a SNORW on 220 nm silicon-on-insulator platform is discussed for optical sensing of gases. Analysis and interdependence of decisive parameters of SNORW are evaluated. Overall performance of SNORW in terms of power confinement, waveguide sensitivity and compactness is compared with that of conventional slot and ridge slot waveguides. Exploratory results on SNORW application as methane sensor are also discussed.

6.3 DESIGN AND OPTIMIZATION

The silicon rectangular waveguide is shown in Fig. 6.1(a). It consists of an array of silicon nanowire of 10 nm diameter (*d*) and 3 nm spacing (*g*) on silicon oxide layer. The height (*h*) of silicon nanowire is kept fixed as 220 nm in silicon wafer technology. The process of fabrication of such a waveguide is given in detail in Reference (Yan, Gargas and Yang, 2009). Total width (*w*) of SNORW for number of nanowires (*k*) in a row can be written as

$$w = (k - 1) \cdot g + k \cdot d \qquad ...(6.1)$$

To compare the properties of SNORW as sensing element with recently reported conventional slot and ridge slot waveguide (Zengzhi *et al.*, 2015) having width of 500 nm, the parameters are taken as $k = 33$, $g = 5$ nm and $d = 10$ nm for illustrative purpose in the present model, which from Eq. 6.1 yields 490 nm as overall width *w*.

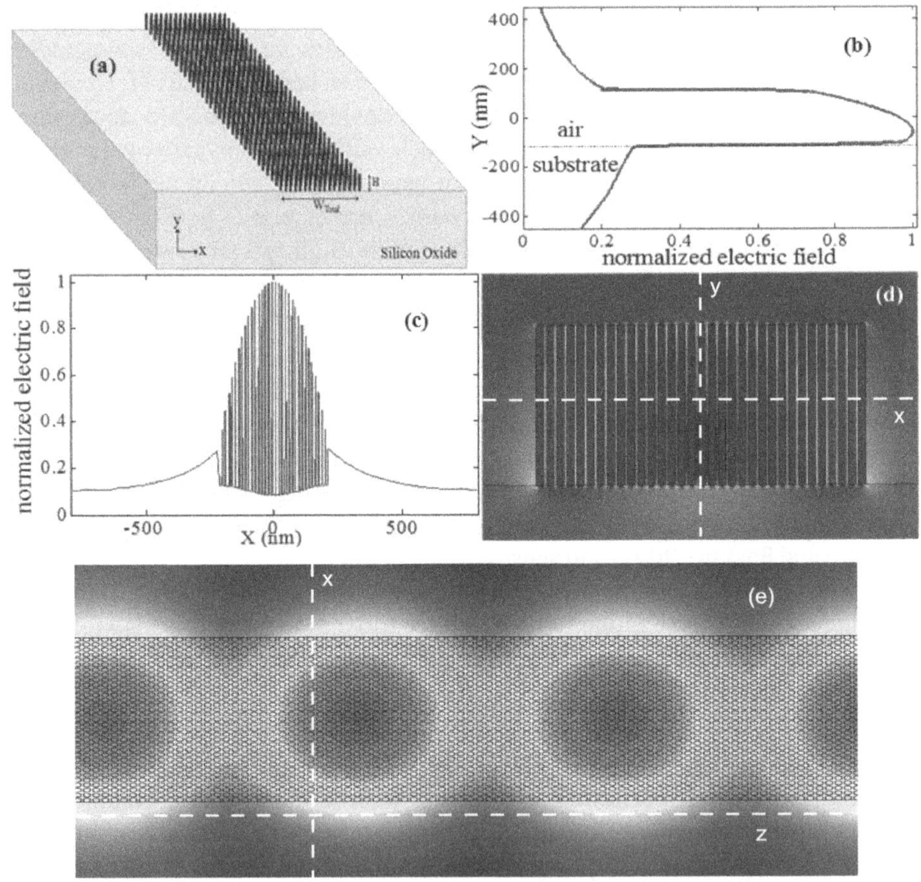

Fig. 6.1. SNORW of h = 220 nm, w = 426 nm. (a) Schematic; (b) Normalized $E_x(0,y)$; (c) Normalized $E_x(x,0)$; (d) Front view of SNORW, $E_x(x,y)$; (e) Top view of SNORW.

The normalized electric field along *y*-axis is shown in Fig. 6.1(b). It can be observed that there is a discontinuity of normalized electric field at cladding and substrate boundary, which is one of the reason for its enhanced confinement in the silicon nanoslot waveguide as compared to conventional and ridge slot waveguides (Zengzhi *et al.*, 2015). The gaps between adjacent silicon nanowires behave as a nanoslot and collectively support fundamental quasi-TE mode. The normalized electric field distribution generated along *x*-axis at a certain height, obtained from finite element method based COMSOL simulation are shown in Fig. 6.1(c), and corresponding two-dimensional field distribution for front and top views are shown in Fig. 6.1(d) and 6.1(e) respectively. A non-uniform meshing of triangular elements is created through using quadratic Lagrange functions. The maximum mesh element size in constant RI region is selected to be 54 nm while minimum mesh size is 0.1 nm along the different materials boundaries. Communication wavelength of 1550 nm is selected as the operating wavelength.

Confinement factor is an essential parameter that is related to the performance of waveguides which act as the sensors, as discussed previously with the help of Eq. 3.1. Fig. 6.2 shows CF which is obtained for SNORW with $g = 5$ nm as a function of RI of the cladding. The RI of silicon oxide substrate is 1.444 and RI of silicon is 3.4752, calculated from Eq. 3.9 and Eq. 3.10 at 1550 nm wavelength. For application as gas sensor, the range of RI is from 1.0 to 1.45, which covers most of important gases including air at infrared optical frequency (Zengzhi *et al.*, 2015). In Fig. 6.2, waveguide confinement factor (Γ_w) > slot confinement factor (Γ_s) because

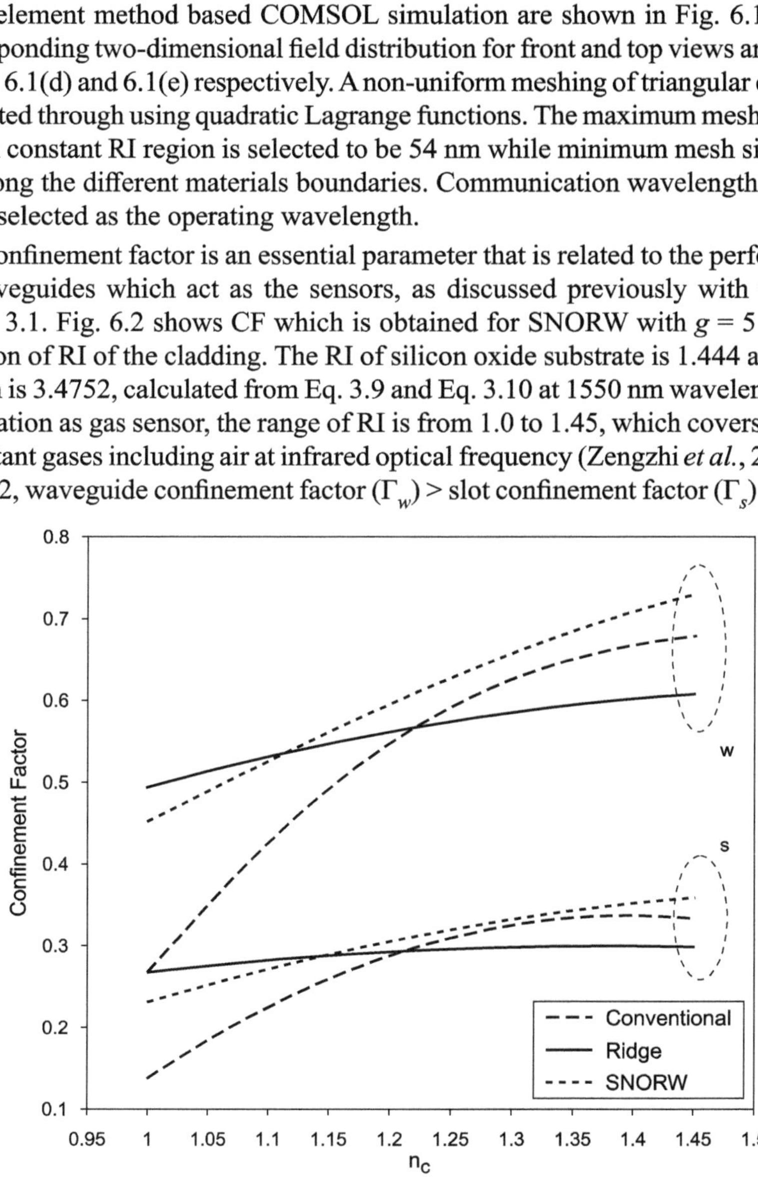

Fig. 6.2. CF versus RI of cladding.

field confinement in the waveguide region is more than that of slot for all values of cladding refractive index. From Fig. 6.2, it can be observed that CF obtained for SNORW are higher than those obtained for recently reported slot and ridge (Dell'Olio and Passaro, 2007; Zengzhi *et al.*, 2015) waveguides when the cladding RI is in the range of 1.15 to 1.45, but it is lower in range 1 to 1.15. Total width of waveguide is 500 nm for ridge and conventional slot, and 490 nm for SNORW. SNORW has the potential to function as an efficient sensor with dimension comparable to waveguides reported in Ref. (Dell'Olio and Passaro, 2007; Zengzhi *et al.*, 2015).

Next, to further optimize the geometrical parameters of SNORW, the CFs are obtained for different values of gap between adjacent nanowires and shown in Fig. 6.3. Calculations are carried out for 3 nm, 4 nm and 5 nm gaps between nanowires.

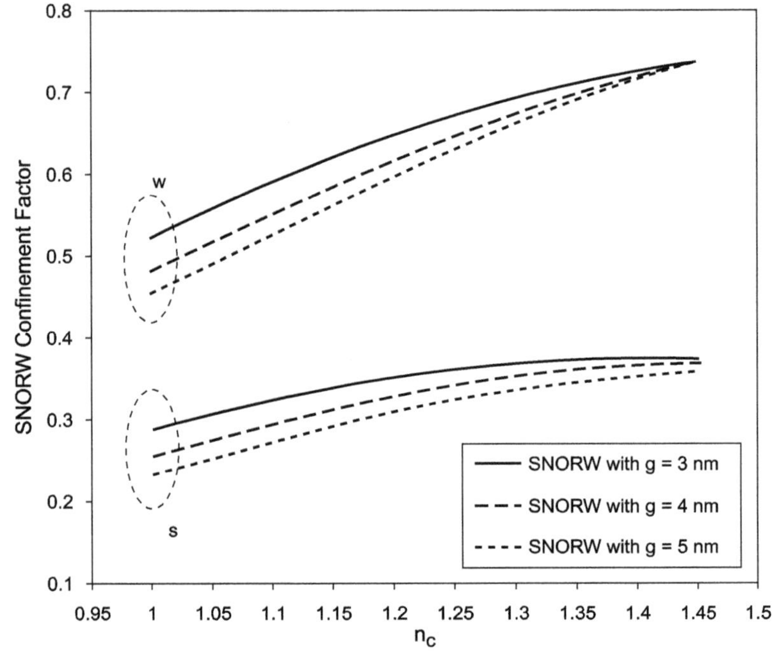

Fig. 6.3. CF of SNORW versus RI of cladding (n_c) with h = 220 nm, d = 10 nm for different nano-slot gapping among nanowires (g = 3, 4 and 5 nm).

It is observed that as the gap between nanowires decreases, the CF in nanoslots and waveguide region increases. Thus, it can be inferred from Fig. 6.3 that SNORW with g = 3 nm has higher power confinement than g = 5 nm and reported values (Dell'Olio and Passaro, 2007; Zengzhi *et al.*, 2015) for the whole RI analysis range of 1 to 1.45. For further calculations, g = 3 nm is used for optimum performance in the whole range of RI of cladding from 1 to 1.45.

6.4 WAVEGUIDE SENSITIVITY

Waveguide Sensitivity for sensing is defined as the ratio of infinitesimal change in effective RI of waveguide to the infinitesimal change in RI of cladding. The sensitivity of a waveguide towards minimal change in RI of cladding due to its exposure to external gases is another most important parameter to make an efficient gas-sensing device. Waveguide sensitivity (S_h) for a homogeneous sensing, is defined as

$$S_h = \frac{\Delta n_{\text{eff}}}{\Delta n_c} \qquad ...(6.2)$$

Where, Δn_c is the change in RI of cladding, which can be the gases to be detected and Δn_{eff} is the change in effective mode index for the propagation of dominant quasi-TE fundamental mode. In this section, waveguide sensitivity for conventional slot waveguide of $w = 200$ nm, $h = 220$ nm and $s = 100$ nm is calculated. Ridge slot waveguide used for sensitivity calculation has the same dimension as of conventional slot waveguide with un-etched height of 20 nm (Zengzhi *et al.*, 2015). The results are compared with optimized SNORW with $g = 3$ nm corresponding to $k = 33$ and $d = 10$ nm. The total width of waveguide is 500 nm for ridge and conventional slot, and 426 nm for SNORW is taken for sensitivity comparison. For variation in RI from 1 to 1.45, waveguide sensitivity of all three waveguides are shown in Fig. 6.4. It is pertinent that SNORW is more sensitive to changes in cladding RI than the conventional and ridge slot waveguides.

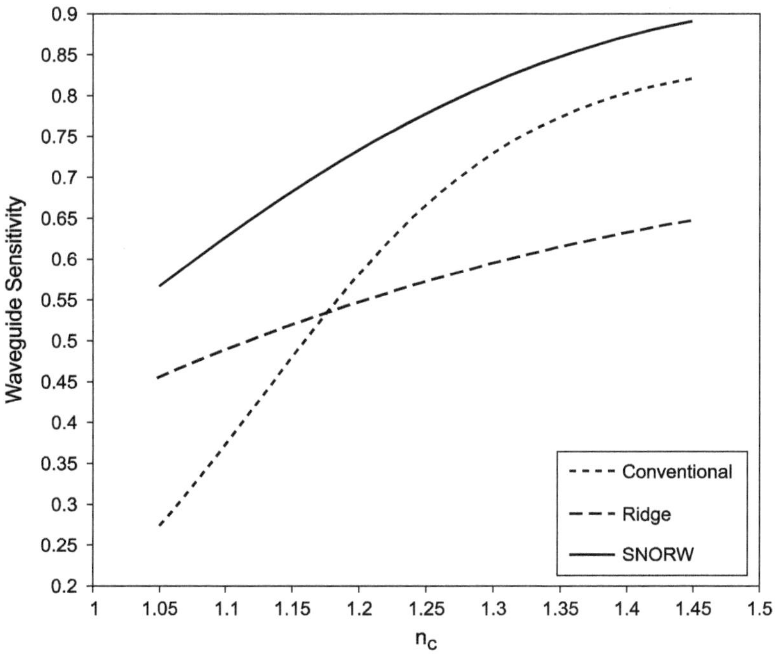

Fig. 6.4. Waveguide sensitivity versus RI of cladding.

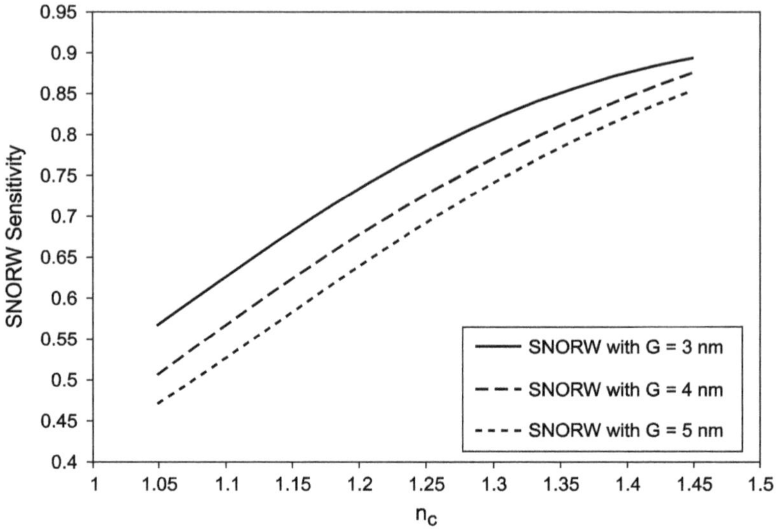

Fig. 6.5. Waveguide sensitivity of SNORW dependence on n_c with h = 220 nm, d = 10 nm for different nanoslot gapping among nanowires (g = 3, 4 and 5 nm).

Further, in Fig. 6.5, it is shown that, as value of g increases sensitivity decreases. Hence, the choice of g = 3 nm based on CF calculation is valid for waveguide sensitivity as well.

6.5 COMPACTNESS

The size of a waveguide is an important parameter to realize multiple sensing of elements in SOI. Total width of SNORW is represented in Eq. 6.1. For conventional and ridge slot waveguides, total width of waveguide can be written as

$$W_{T.\text{Conventional}} = W_{T.\text{Ridge}} = (2 \cdot w_s) + s \qquad ...(6.3)$$

where, 'w_s' is the width of each silicon slab and s is the slot width. For w_s = 200 nm and s = 100 nm, total widths of both conventional and ridge slot waveguides are 500 nm. For SNORW, values of k and d is taken as 33 and 10 nm respectively and values of width 'w' are obtained as 426 nm, 458 nm and 490 nm by using Eq. 6.1 for g = 3 nm, 4 nm and 5 nm, respectively. Hence, SNORW can be fabricated with better performance than the two waveguides as reported in (Zengzhi *et al.*, 2015), by reducing the size by 14.8% for g = 3 nm.

6.6 GAS SENSING

In this section, a preliminary analysis is performed to evaluate SNORW sensor that detects methane gas in coalmines. Leakage of methane gas is not an instantaneous act and it is to be detected for a very low percentage by volume in order to avoid any calamity. The presence of methane in air will gradually modify the RI of gas-air mixture which will result in alteration of the atmospheric refractive index.

The RI of methane-air mixture with different percentage by volume is obtained by Lorentz-Lorenz mixing rule (Tasic *et al.*, 1992) and is given in Eq. 6.4 and rearranged into more practical form in Eq. 6.5.

$$\frac{n_{ab}^2-1}{n_{ab}^2+2} = \phi_a\left(\frac{n_a^2-1}{n_a^2+2}\right)+\phi_b\left(\frac{n_b^2-1}{n_b^2+2}\right) \qquad ...(6.4)$$

$$n_{ab} = \sqrt{\frac{(n_a^2+2)(n_b^2+2)+2\cdot(\phi_a(n_a^2-1)(n_b^2+2)+\phi_b(n_a^2+2)(n_b^2-1))}{(n_a^2+2)(n_b^2+2)-(\phi_a(n_a^2-1)(n_b^2+2)+\phi_b(n_a^2+2)(n_b^2-1))}} \qquad ...(6.5)$$

where $\phi_i = w_i\rho_{ab}/\rho_i$, $w_i = m_i/(m_a + m_b)$, and $i = a$ and b (Tasic *et al.*, 1992). ρ is density, m is molar mass, w is mass fraction, ϕ is volume fraction and n is refractive index. The possibility of detection of different percentages by volume of methane–air mixture is reflected by calculating the guided power for effective changes in RI of cladding due to the exposure of methane. Fig. 6.6 shows that even for 0.05% change in volume from pure air to methane–air mixture, normalized power decreases by 5% which can be detected by low cost optical power meters. Normalized power (P_{NP}) in cladding region is defined as

$$P_{NP} = \left|\frac{E_{ab}}{E_a}\right|^2 \qquad ...(6.6)$$

Here, E_{ab} is the electric field in nanoslot region for the mixture of air and methane, E_a is the electric field in nanoslot for air at STP, and a, b are the representation for air and methane respectively. For higher values of methane–air mixture, the normalized power level goes substantially up to 50% when air is

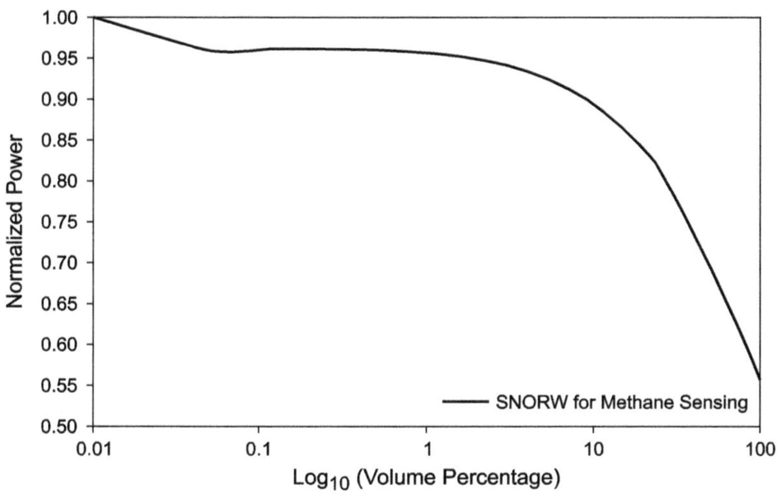

Fig. 6.6. Normalized power in nano-slots versus percentage volume in logarithmic scale of SNORW for methane sensing with h = 220 nm, d = 10 nm, g = 3 nm.

completely replaced by methane. However, to avoid any hazard in coalmines, methane–air mixture should be within 5% (Leonardis, 2012) for which SNORW will be very effective.

6.7 SUMMARY OF GAS SENSING

Silicon nanowire optical rectangular waveguide with enhanced characteristics for sensing application is discussed. Characteristic properties like confinement factor, waveguide sensitivity, compactness and application as methane sensor are successfully studied by applying finite element method modal analysis. By appropriately designing the SNORW parameters, its size can be reduced by 14.8% compared to that of conventional and ridge slot waveguides with enhancement in CF and waveguide sensitivity. For refractive index in the range of 1 to 1.45, average confinement factor of SNORW is (G_s =) 34% for slot and (G_w =) 65% for waveguide regions, which is higher than that of ridge (29% for slot and 56% for waveguide regions) and conventional (27% for slot and 53% for waveguide regions) slot waveguides. The waveguide sensitivity, which is another important parameter, is also higher for SNORW with respect to other two waveguides. For RI = 1.0, the sensitivity of SNORW (51%) is relatively higher than that of ridge and conventional slot waveguides by 11.2% and 29.4%, respectively. Moreover, in Ref. (Dell'Olio and Passaro, 2007; Zengzhi *et al.*, 2015), the RI of cladding is changed assuming air is replaced 100% in volume by the gas. However, it is important to measure the small percentage by volume as the leakage of gas is disseminated inside the atmosphere of colliery. In this work, for application as methane gas sensor, it is shown that the SNORW can detect less than 5% by volume of methane-air mixture that is the limit to avoid explosion in coalmines. The dimensions of SNORW used here, are on lower side which is expected to be further improvised with advancement in silicon wire fabrication technology.

❑❑❑

7 Blood Glucose Sensing Applications

In this chapter, a silicon nanowires based optical waveguide for sensing application is demonstrated with analysis. SOI based PIC structures using refractive index sensing is done for various waveguide structures. The chapter gives detailed analyses of SNORW in blood glucose sensing application. Minimal invasive approach is being adopted here by taking blood as sample material. Measurand includes a mixture of blood sample, NaF and EDTA in specific fractions. The characteristic properties of the sensor have been evaluated through FEM analysis and these properties include electric field variation, confinement factor, change in effective refractive index of the waveguide, waveguide sensitivity, mode propagation area and relative intensity response.

7.1 ADVANCEMENT IN PHOTONIC SENSING

Recent technological advancement in photonics industry has led to a resurgence of interest in optical glucose sensing by components of Photonic Integrated Circuit. In this section, a new glucose sensor based on optical waveguide is discussed. Nowadays, photonic sensing technologies satisfy a goal of making compact and highly sensitive sensors capable to detect a very small volume of a specific bio-chemical sample and they are fast and accurate too. The monitoring of blood glucose level helps in effective diagnosis and adequate treatment of hypo-glycaemia and hyper-glycaemia.

Minimal-invasive approach is used here for measurement of glucose concentration. To evaluate the characteristics of this sensor, Ethylene-diamine-tetra-acetic acid (EDTA) as an anticoagulant, sodium fluoride as preservative and blood sample as sensing measurand are added in the fraction of 0.8, 0.1 and 0.1, respectively. As the glucose concentration in blood sample varies, the refractive index of blood changes, accordingly the refractive index of the solution with 10% blood also gets altered, which in turn deviates the response of biosensor. The prediction of glucose level is affirmed by taking this solution as a cladding measurand in the next step of sensing process. SNORW based biosensor is been discussed for the first time to detect blood glucose level. It works on the principle of detecting changes in refractive index for various concentrations of glucose level ranging from 10 mg/dl to 200 mg/dl.

DOI: 10.1201/9781003783589-7

In SNORW, the gaps between adjacent silicon wires behave as a nanoslot and collectively support the fundamental quasi-TE mode. Considering the fabrication technology of 220 nm silicon height, first a typical configuration of SNORW is optimized to provide physical compactness and high power confinement. The mode field distribution for SNORW as well as slot waveguide are obtained by using Finite Element Method. This minimal-invasive glucose sensing is performed in comparison of SNORW with conventional slot waveguide. For better comparison, slot waveguide is taken with comparable parametric waveguide width to that of SNORW. Characteristic properties like confinement factor, waveguide sensitivity, compactness and application as glucose sensor are successfully studied for SNORW and compared to conventional slot waveguide. By appropriately designing the SNORW and conventional slot waveguide parameters, the sensitivity can be increased by more than 20% in case of SNORW. Also, the power confinement factor of SNORW is more than that of conventional slot waveguide by 10%. The relative intensity response steeply decreases as the concentration of glucose increases. Silicon nanowire waveguide exhibits a sensitivity significantly larger than the previous PIC structures due to its high surface to volume ratio. For 0.5% change in fraction of blood glucose concentration, normalized power decreases by >4% which can be detected by low cost optical power meters. An approach to enhance the difference in power confinement in presence of <1% change in glucose concentration is demonstrated.

7.2 NEED OF BLOOD GLUCOSE SENSOR

Nowadays, biosensors are being widely used in detection of an analyte. Due to recent advancement in photonic technologies, it has been investigated that they have find their ways in optical sensing applications. This technology helps to provide miniaturized, nonelectrical, highly sensitive, fast and accurate biosensors. These photonic biosensors are reliable, easy to handle, have minimum fluctuation in measuring results over time and have subparts which can be easily replaced when needed (McNichols and Coté, 2000). Biosensors, by means of photonic integrated on-chip circuits, have been proposed in recent trends. A widely adopted approach includes biosensors based on rectangular waveguide and slot waveguide (Li *et al.*, 2017). The rectangular waveguides where light is confined in the high refractive indexed region are applicable as biosensors due to their high refractive index contrast and less transmission loss (Mukundan *et al.*, 2009; Kozma *et al.*, 2014). But, these biosensors have limited sensitivity because of weaker overlap between light and sensing material. In contrast to this, adopting a slot waveguide would improve the sensitivity as it has strongly confined and guided light in low refractive index area, which would result into the light field distribution which is rather sensitive to the refractive index of the measurand.

Diabetes mellitus, a very common disease nowadays, in which body either fails to respond to the consumption of glucose level or fails to produce adequate amount of insulin. Diabetic patients can be mainly bifurcated as hyperglycemia and hypoglycemia. Hyperglycemia, which has a glucose level higher as compared to a healthy person, is caused when the human body is not able to utilize glucose for generation of fat and energy, resulting into production of toxic and high level ketone inside the human body. Hence, hyperglycemic patients may face consequences like ketoacidosis, coma or even death (Koschinsky and Heinemann, 2001). On the other hand, hypoglycemia includes low glucose level patients who have insufficient glucose level in their body to generate adequate amount of energy for physical metabolic activities of human body, resulting into weakness and unconsciousness due to low fat level. Thus, monitoring blood glucose level becomes important for effective diagnosis and treatment of hyperglycemia and hypoglycemia patients to supply adequate amount of insulin or medicine.

The glucose level can be detected either by invasive or non-invasive approach (Koschinsky and Heinemann, 2001). Non-invasive technique, unlike the invasive one, does not require interstitial fluid for the detection of glucose level. However, measurement through non-invasive approach is not as precise as invasive approach. Especially minimal invasive approach is advantageous in many aspects as it requires a very small sample for glucose detection. For majority of healthy individuals' normal blood sugar levels are between 72 – 108 mg/dl ('Diagnosis and Classification of Diabetes Mellitus', 2014) during fasting and less than 140 mg/dl 2hrs after meals. Some of the intense researches have been performed glucose detection using various photonic methods. A demonstration of glucose sensing based on monitoring of transmission and absorption of spectrums for rib silicon waveguide was given as a sensing scheme (Ryckeboer *et al.*, 2014). Earlier, slot waveguide chip, based on silicon-on-insulator technology had explained how the optical power guiding property in low refractive index region improves the waveguide sensitivity (Dell'Olio and Passaro, 2007) resulting in much effective sensing operation. In fact, use of this slot waveguide for refractive index sensing purpose is likewise explained in both homogeneous as well as surface material sensing (Li *et al.*, 2017). On the other hand, tapered fibers are used for glucose sensing in which fibers are spliced to work like Mach-Zehnder modulator (Bhardwaj and Singh, 2016). Afterwards, this tapered modulator was immersed in glucose solution and the transmission spectra give the relation with glucose concentration. In the same year, photonic crystal fiber based glucose sensing methods were investigated for hypoglycemia episodes (Yang *et al.*, 2016; An *et al.*, 2017). Meanwhile, some researchers have also tried to sense the level of glucose by correlating it to wavelength shift in ring resonator structure using tunable laser source (Luo *et al.*, 2014). Later, photonic crystal waveguide based glucose sensor were developed in which cubic lattice of circular silicon rods are used for creating defects (Robinson and Dhanlaksmi, 2017).

The aim of this article is to critically discuss the technical aspects of glucose sensing on silicon on insulator platform using PICs. It works on the principle of detecting changes in refractive index for various concentrations of glucose levels ranging from 10 mg/dl to 200 mg/dl. Further, this sensing technique is compared with the conventional slot waveguide. Silicon nanowire waveguide exhibits a sensitivity significantly larger than the previous PIC structures due to its high surface to volume ratio. The method includes minimal-invasive sensing technique for spot glucose measurement used by hypoglycaemia and hyperglycaemia patient to get optimize insulin therapy. The prediction of glucose level is affirmed by taking blood sample as a cladding measurand for sensing process.

We aimed to provide a potential effective method for glucose detection, the structure met with the requirement needed to be satisfied by glucose sensors as: the measured results are immediately available, the frequency of measurements is very high, and these sensors are reliable having high precision. This sensor is capable to detect a very small volume of a specific bio-chemical sample. Optical glucose sensing method solves the calibration problem. The concentration of sample covers the dynamic level of glucose for both healthy and diabetic persons.

7.3 WAVEGUIDE DESIGN AND SENSING METHODOLOGY

Two types of SOI waveguides, one is conventional slot waveguide, and another is SNORW, are taken in this section. In order to compare these two structures for glucose sensing, the upper cover region of sensing waveguides is kept empty as shown in Fig. 7.1(a) and 7.1(b). The coupling at the input and output end of these sensing structures are conventional rectangular waveguide. Silicon nanowire based optical waveguide is sandwiched between rectangular waveguide at input and output region for better sensing because of the large value of surface to

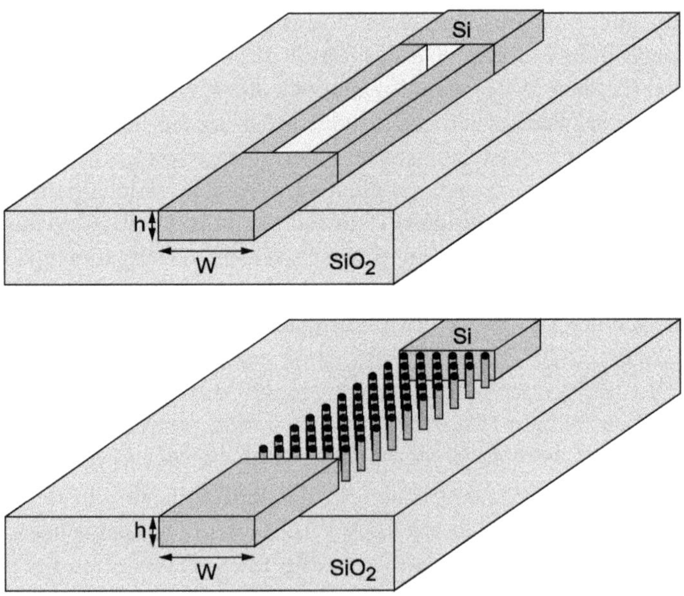

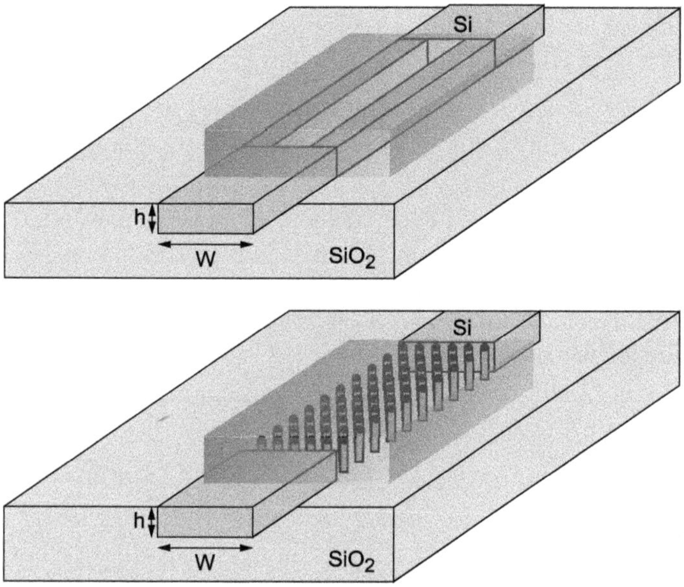

Fig. 7.1. Schematic of SNORW glucose sensor.

volume ratio in SNORW. The structural arrangement of these sensing waveguides with measurand at cover region is shown in Fig. 7.1(c) and 7.1(d). The width of rectangular waveguide is taken as '*w*', height of rectangular waveguide as '*h*', diameter of single nanowire as '*d*', gapping between nanowires as '*g*', and the number of nanowires in a row is taken as '*k*'.

Firstly, air cladding material is considered to perform all the characteristic analysis. Afterwards, cladding material is replaced by a sample mixture which contains 0.1 fraction of blood sample as sensing measurand, 0.1 fraction of sodium fluoride (NaF) as preservative and 0.8 fraction of EDTA as an anticoagulant. Homogeneous sensing is performed here and SiO_2 is taken as cover as well as substrate material. The refractive index values for all the composition materials are taken at 1550 nm wavelength as 3.4752, 1.4444 and 1 for n_{Si}, n_{SiO_2}, and n_{air}, respectively. Refractive index of blood sample is been calculated using the formula (Yeh, 2008).

$$n_{blood} = (0.011889 \cdot c) + 1.33230545 \qquad ...(7.1)$$

Where '*c*' is the concentration (mg/dl) of glucose in blood. The refractive index of whole composition for different concentration of glucose is been acquired by using Lorentz-lorenz mixing rule (Tasic *et al.*, 1992). We have derived Lorentz-lorenz mixing rule for a mixture consisting of three different materials with different proportion. The derived equation is given as:

$$\frac{n_m^2 - 1}{n_m^2 + 2} = \phi_1\left(\frac{n_1^2 - 1}{n_1^2 + 2}\right) + \phi_2\left(\frac{n_2^2 - 1}{n_2^2 + 2}\right) + \phi_3\left(\frac{n_3^2 - 1}{n_3^2 + 2}\right) \qquad ...(7.2)$$

$$n_m = \sqrt{\frac{1+2A+2B+2C}{1-A-B-C}} \qquad ...(7.3)$$

$$A = \phi_1\left(\frac{n_1^2-1}{n_1^2+2}\right), B = \phi_2\left(\frac{n_2^2-1}{n_2^2+2}\right) \text{ and } C = \phi_3\left(\frac{n_3^2-1}{n_3^2+2}\right)$$

Where, n_m is refractive index of the mixture, n_1 is refractive index of NaF, n_2 is refractive index of EDTA, n_3 is refractive index of blood; ϕ_1, ϕ_2 and ϕ_3 are fractions of NaF, EDTA and blood in the mixture, respectively. The calculated values of refractive index of n_1 and n_2 at 1550 nm is given as n_1 = 1.3194 (Li, 1976) and n_2 = 1.363.

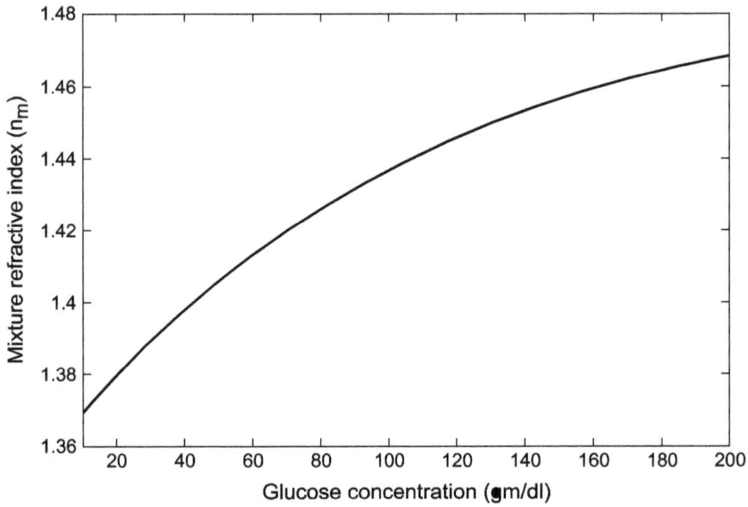

Fig. 7.2. Mixture refractive index with variation in blood glucose concentration.

The value of $n_3 = n_{blood}$ is taken for different concentration of glucose ranging from 10 mg/dl to 200 mg/dl. The refractive index of the mixture has been calculated with the help of Eq. 7.3 and shown in Fig. 7.2. Fraction of NaF, EDTA and blood in mixture are 0.1, 0.8 and 0.1, respectively. Increase of glucose concentration in blood in mixture, results in increase in its refractive index.

7.4 PARAMETRIC ANALYSIS FOR GLUCOSE SENSING

In this sensing analysis of SNORW, parametric values of *h*, *d*, *g* and *k* for waveguide optimum performance are taken as 220 nm, 10 nm, 5 nm and 33, respectively, which in turn result to total waveguide width of SNORW as 490 nm. The cladding portion of this SNORW is taken as air, which will then be further replaced by blood mixture measurand for glucose sensing. Power in the SNORW structure is coupled through a silicon ridge waveguide at both input and output side. For better understanding of the sensing behavior of this SNORW structure, another waveguide structure i.e., slot waveguide is used for comparison. Slot waveguide with slot width of 100 nm and silicon brick width of 200 nm is taken for most compelling evidence of overall waveguide size. Total waveguide width of slot waveguide is 500 nm which is comparable to SNORW and likewise coupled with ridge waveguide of w = 500 nm. Also, the cladding portion of slot waveguide is replaced by measurand in the same fashion as of SNORW sensing structure. Further, analysis is performed with Finite Element Method (FEM) for efficient analysis of designed structure, in which free triangular mesh is taken with grid size declaration as: minimum element size is 0.01 nm, maximum element size is 1 nm, resolution of narrow region is 100, and element growth rate is 1.1. Considering all these, the different key characteristic properties have been observed and analyzed.

7.4.1 Mode Field Profile

Electric field along x-cut is obtained and plotted simultaneously for both, air cladding and glucose cladding. Fig. 7.3(a) shows the mode confinement, that can be seen much stronger at the center and weaker towards the edges. The peak of confined normalized electric field goes downward when the glucose cladding is used as measurand. This analysis being performed uses the glucose concentration as 70 mg/dl.

Electric field along y-cut is observed and plotted in both the cladding conditions as shown in Fig. 7.3(b). It becomes maximum in the gap between wires and decreases steeply in the substrate and cover region. The confinement of E-field decreases with decrease in glucose concentration of cladding. Considering the case of air filled in the gaps between nanowires, a peak of electric field is found near the substrate

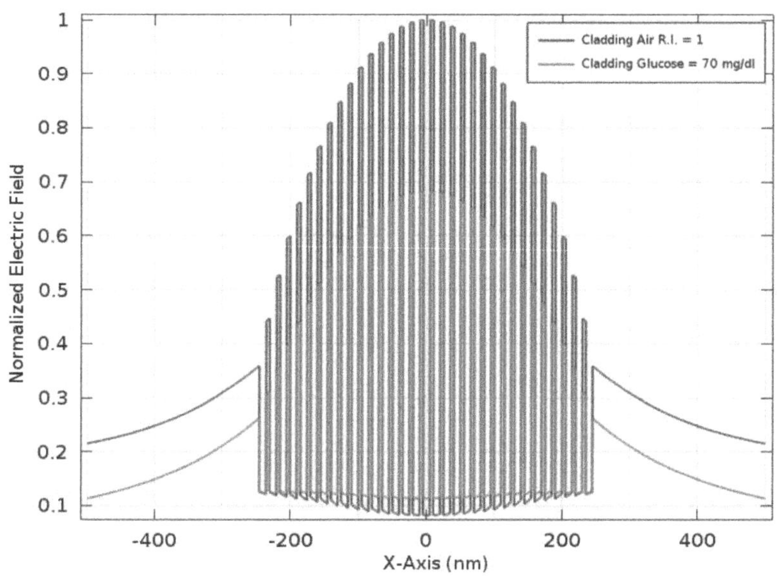

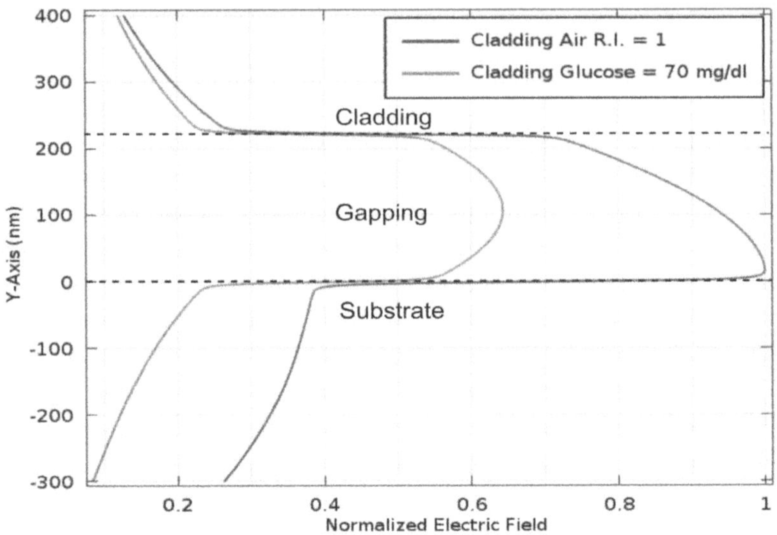

Fig. 7.3. Mode field (a) *x*-cut; and (b) *y*-cut; distribution profile of SNORW when cladding is measurand or air.

interface. However, in the case of glucose concentration of 70 mg/dl filled in between, the electric field spreads to maximum in the middle of gapping between these wires.

7.4.2 Effective Refractive Index

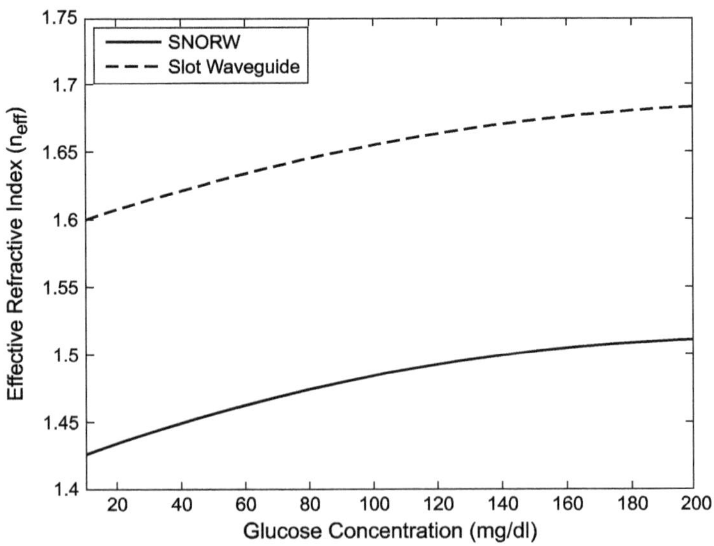

Fig. 7.4. Effective refractive index with variation in glucose concentration.

The variation in effective refractive index (n_{eff}) of silicon nanowire waveguide with different concentration of glucose can be seen in Fig. 7.4. It is increasing with the increase in glucose concentration from 10 mg/dl to 200 mg/dl. A slot waveguide followed the same pattern of variation. Moreover, effective refractive index of silicon-based waveguide mainly depends on the RI of silicon material, and silicon is found to have higher RI in nature than other compositions. Since, the volume composition of silicon material is high in slot waveguide than that of SNORW, thus, the overall effective RI of slot waveguide is 12% higher than that of the SNORW waveguide for all glucose concentration ranging from 10 mg/dl to 200 mg/dl.

7.4.3 Effective Mode Area

The effective mode area represents the study of mode propagation area of the waveguide. In case of SNORW waveguide, it decreases as the glucose concentration increases from 10 mg/dl to 200 mg/dl. The value of effective modal area is high for low glucose concentration in blood sample. However, upon using slot waveguide for the glucose sensing purpose, the mode area slowly tends on increasing with the increase in glucose concentration. The effective area of the waveguide can be calculated by using Eq. 3.2 and shown in Fig. 7.5.

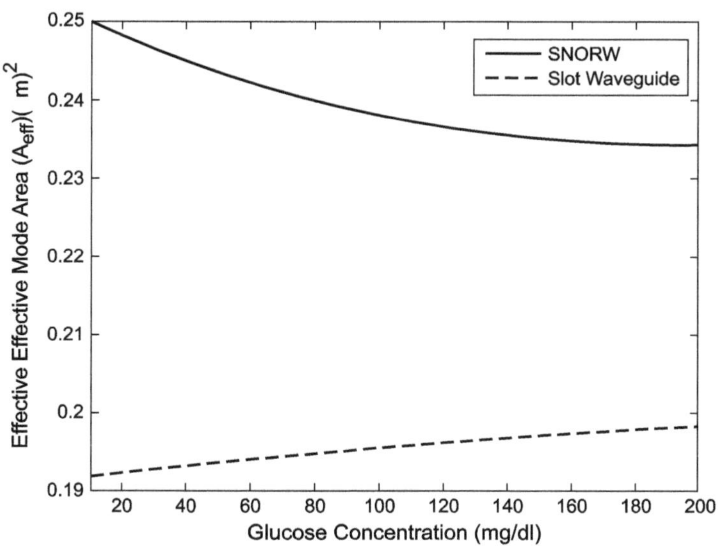

Fig. 7.5. Relationship between effective mode area and glucose concentration.

7.4.4 Confinement Factor

Power confinement factor in any particular area can be obtained by taking the ratio of power in the desired region to the power in the total region. The confinement factor represents the total power confined in the particular area. It can be calculated using Eq. 3.1. In case of slot waveguide, the light confinement is stronger when the index contrast between the silicon and the measurand is larger (Kargar and Chao, 2011). Larger percentage of light inside the low-index material, facilitates stronger evanescent field to interrogate the analyte. In principle, larger index contrast offers higher confinement factor. Therefore, silicon slot waveguide with high glucose concentration has low confinement factor inside slot region as shown in Fig. 7.6 with dotted lines. However, using SNORW instead of silicon slot waveguide, the confinement factor increases with increase in glucose concentration.

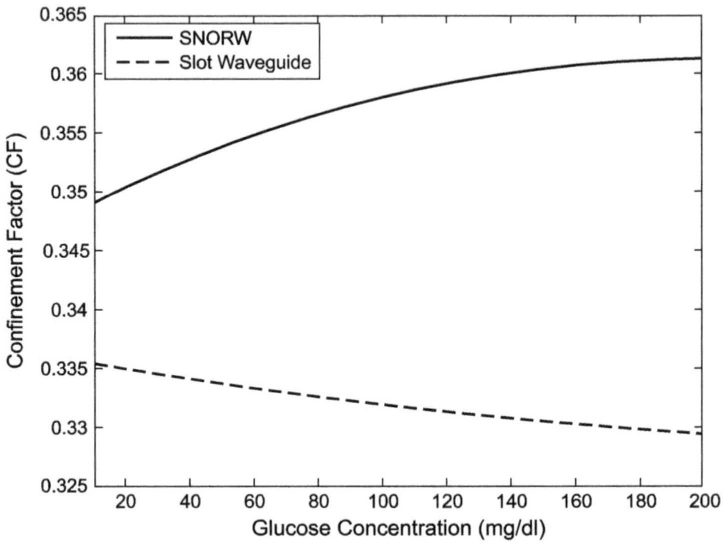

Fig. 7.6. Variation of confinement factor with glucose concentration.

7.4.5 Sensitivity

The sensitivity is strongly related to the amount of the evanescent wave or the percentage of light inside cladding region of waveguide, where the analytes are available. The sensitivity of a SNORW or slot waveguide sensor can be expressed with waveguide sensitivity (S_h). The waveguide sensitivity is characterized as the ratio of change in effective index to the change in analyte's refractive index. Waveguide sensitivity is related to the amplitude of evanescent wave and the percentage of light inside low-indexed material for analyte interrogation. Slot and SNORW waveguides due to their unique characteristic, confine the majority of light within low refractive index slot, resulting into higher waveguide sensitivity as compared to other conventional waveguides, for example, ridge waveguide. Sensitivity is one of the key parameters to evaluate sensor performance. SNORW is advantageous because of its high sensitivity and can preferably work as a good sensor. The waveguide sensitivity is expressed in Eq. 6.2. and it is plotted against glucose concentration for both SNORW and slot waveguide as shown in Fig. 7.7.

It is observed that the device becomes more sensitive as concentration of glucose increases from 10 mg/dl to 200 mg/dl. A silicon slot waveguide also provides same variation in the plot, but one can clearly observe that the exceeding variation in sensitivity per glucose increment can be better depicted in the case of SNORW, which is an essential attribute for a good sensor.

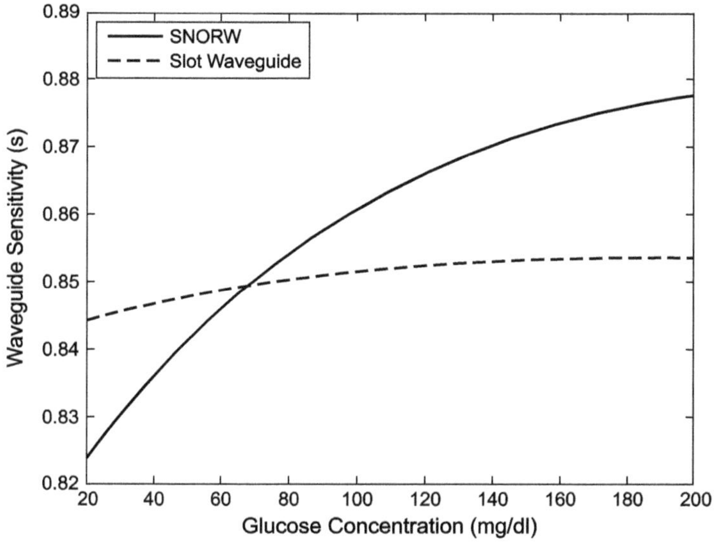

Fig. 7.7. Waveguide sensitivity for different glucose concentration.

7.4.6 Through-port Intensity

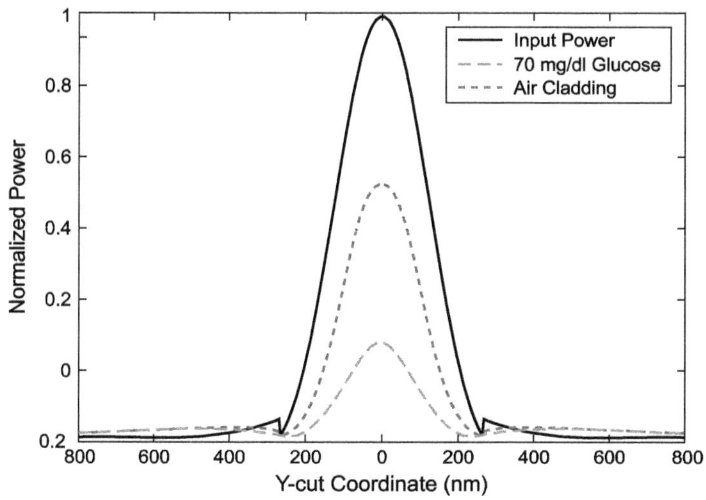

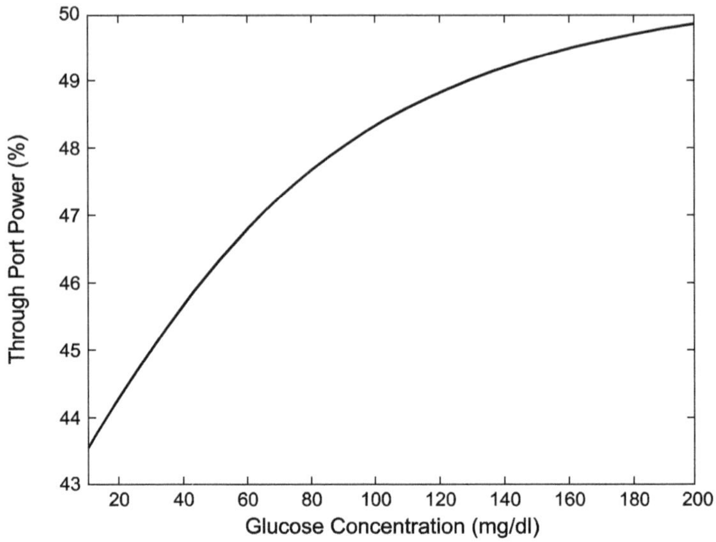

Fig. 7.8. (a) Normalized y-cut (along ridge waveguide width) distribution profile of SNORW; (b) Through port power for different glucose concentration.

The output of SNORW is coupled with ridge waveguide having width same as total waveguide width of SNORW or slot waveguide. For this, SNORW with $k = 33$, $d = 10$ nm and $g = 5$ nm having total waveguide width of 490 nm and a corresponding ridge waveguide width of 490 nm is chosen. To sense the propagating field inside SNORW configuration, input and output are coupled with the ridge waveguide. The power is measured along y-cut and plotted in Fig. 7.8. The mapped out normalized power profiles in the first graph corresponds to the input power, glucose with 70 mg/dl concentration and air cladding, which is represented via blue, green and red color, respectively. The input power when passed through SNORW and again coupled with ridge waveguide, decreases the power level according to the measurand refractive index. Fig. 7.8(b) shows through-port coupled power inside ridge waveguide for 10 mg/dl to 200 mg/dl glucose concentration. In case of SNORW, the light confinement becomes stronger as the refractive index of mixture inside cladding region of SNORW increases. Therefore, SNORW with high glucose concentration is having high through-port power.

7.5 SUMMARY OF BLOOD GLUCOSE SENSING

The biosensor is realized to potentially deliver an effective optical glucose sensing on SOI platform using photonic integrated circuits, which is highly reliable and remarkably precise. Silicon nanowire exhibits a sensitivity that is significantly larger than the previous photonic integrated structures, due to its high surface to volume ratio. The method used in this section includes minimal invasive sensing technique for spot glucose measurement used by hyperglycemic and hypoglycemic patients to get the appropriate insulin therapy. The structure is designed mainly to operate on 1550 nm wavelength. This sensor is capable to detect very small volume of a specific bio-chemical sample. The blood sample is taken as a measurand whose concentration covers the sensing of dynamic level of glucose for both healthy and diabetic persons. Optical glucose sensing solves the calibration problem.

❑❑❑

8 DNA Hybridization Sensing Application

In this chapter, a silicon nanowire based optical waveguide for sensing application is demonstrated with analysis. SOI based PIC structures using refractive index sensing is done for waveguide structures. The chapter provides a thorough investigation for optical sensing of DNA hybridization using SNORW. Characteristic properties like power confinement, mode field intensity, normalized power and waveguide sensitivity are successfully studied for optimized dimensions. The change in refractive index during the hybridization process is identified for the different SOI fabrication technologies. Renaturation and separation through power confinement when different percentages of dsDNA and ssDNA coexist in a sample are evaluated.

8.1 NEED OF DNA HYBRIDIZATION SENSING

Due to recent technological advancement in photonic integrated circuits, silicon waveguides have been used in different bio-sensing applications. Recently, label-free bio-sensing has gained considerable attention by researchers to overcome the difficulties associated with labeled bio-sensing (Ciminelli *et al.*, 2013; Dell'Olio *et al.*, 2015). In bio-sensing, the development of an effective DNA-sensing system is extremely important, in which technologies for DNA hybridization are currently under intense investigation by researchers worldwide. The complicated formalization process involved in DNA hybridization can be overcome by using label free detection involving linker layer rather than conventional targeted DNA sequence (Barozzi *et al.*, 2017). In DNA hybridization process, breaking of hydrogen bonds between complementary base pairs of dsDNA results in ssDNA; whereas ssDNA of two species when allowed to renaturate, yields into hybrid DNA. Similar strands join together and form complementary base pairs by reforming hydrogen bonds. The degree of similarity between two species can be estimated by DNA hybridization percentage.

In recent years, optical biosensors based on dielectric and plasmonic waveguides (Raghuwanshi, Kumar and Athokpam, 2016; Dell'Olio, Conteduca, De Palo, 2017; Yuan and Fang, 2017) are attracting attention of many researchers worldwide. Research has shown that optical sensors with gold layer in plasmonic waveguide are much more effective to detect analytes as compared to other plasmonic biosensors (Gelfand *et al.*, 2011). However, optical plasmonic waveguide or hybrid plasmonic

DOI: 10.1201/9781003783589-8

waveguide have high loss and low propagation length as compared to dielectric waveguide. DNA hybridization can be realized by using both plasmonic as well as dielectric waveguide (Khorasaninejad *et al.*, 2011; Dar *et al.*, 2012; Viphavakit *et al.*, 2015). Silicon waveguide with its inherent advantage of low material cost can be applied to DNA hybridization sensing using horizontal (Viphavakit *et al.*, 2015) and vertical slot waveguides (Dar *et al.*, 2012), fabricated on Silicon-on-Insulator substrate. In slot waveguides, confinement of power is in low index region with high sensitivity have been reported for DNA hybridization (Dar *et al.*, 2012; Viphavakit *et al.*, 2015). Horizontal slot waveguides (Viphavakit *et al.*, 2015) have high surface to volume ratio than vertical slot waveguides (Dar *et al.*, 2012), hence horizontal slot waveguide is better DNA hybridization sensing device as compared to vertical slot waveguide.

Current literature has acknowledged that an arrangement of silicon nanowires array on silicon oxide substrate can work as an optical waveguide and follow the same principle of guidance as of slot waveguide. In such silicon nanowire optical rectangular waveguide, the light is highly confined in low refractive indexed nanoslots between silicon wires, resulting in strong interaction between light and sensing material. Moreover, silicon nanowire-based waveguides have a very high surface to volume ratio. Thus, the power confining ability and field distribution are highly sensitive (Khorasaninejad and Saini, 2011) to the slight change in refractive index of measurand. Silicon nanowire optical rectangular waveguide shows advantage of high confinement factor, high sensitivity, small size, easily integrable, fast response and highly selective with low power consumption.

Silicon nanowire optical rectangular waveguide-based biosensor is used to detect deoxyribonucleic acid (DNA) hybridization (Singh and Priye, 2018; Singh *et al.*, 2019). It works on the principle of detecting the change in refractive index during the hybridization process when double strand DNA splits into single strands of DNA or when two single strands renaturate into a double strand DNA. Considering the fabrication of height 220 nm and 330 nm of silicon, first, a typical configuration of SNORW is optimized to provide physical compactness and high-power confinement. Next, parameters characterizing link layer and DNA are added and change in power confinement are ascertained when double strand DNA (dsDNA) splits into two single strand DNA (ssDNA) or two ssDNA renaturate to form a dsDNA. A means to enhance the difference in power confinement in presence of 100% ssDNA and 100% dsDNA is also demonstrated. As the splitting/renaturation of DNAs is a slow process, power confinement is also evaluated when different percentages of dsDNA and ssDNA coexist.

8.2 PHYSICAL CONFIGURATION AND PRINCIPLE

In Silicon Nanowire Optical Rectangular Waveguide, silicon wires are mainly processed into cylindrical shaped through the diffusion of spherical shaped *Au* nano-particles with silicon material. As a result, vertically grown SiNW having cylindrical 2-dimensional periodic array pattern with circular cross-section is formed, as shown in Fig. 8.1(a). SNORW is developed in accordance to SOI technology, in which fused silica (SiO_2) as a substrate is used. The silicon wires have cross sectional dimension of few nanometers and are named as nanowires or nanorods. These nanorods collectively form a waveguide and support optical wave propagation with dominant Quasi-TE mode. Gaps between these nanowires hold field intensity and mode distribution inside lower refractive index material.

Here, SNORW is utilized to detect DNA hybridization, for which silicon nanowires are coated with Silanes, which acts as a linker/glue (Malinsky *et al.*, 2001; Dar *et al.*, 2012) between silicon and DNA. DNA layer of 8 nm fixed width is glued over linker layer of 1 nm width (Dar *et al.*, 2012), as shown in Fig. 8.1(b) and Fig. 8.1(c), where *h*, *g* and *d* represents height, gap and diameter of silicon rods respectively with *k* number of wires in one row of SNORW structure. The entire arrangement of SNORW with linker and DNA is kept inside water (H_2O) as shown in Fig. 8.1(b). The top view of one silicon nanorod is shown in Fig. 8.1(c).

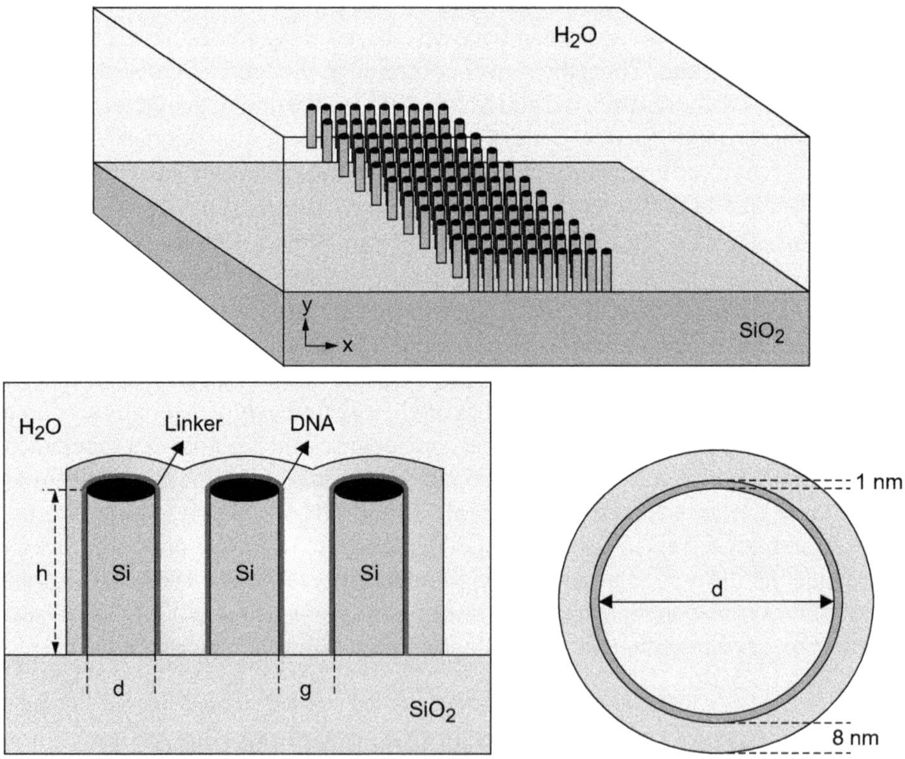

Fig. 8.1. Schematic of SNORW for detection of DNA hybridization. (a) SNORW structure; (b) cross sectional view (*x*-*y* plane) with linker and DNA layer; and (c) top view (*x*-*z* plane) of single Si nanowire.

8.3 POWER CONFINEMENT OF SNORW

To increase the robustness of the SNORW structure, it is essential to calibrate all the parameters at a specific value to obtain maximum efficiency and compactness in the structure. By FEM based optical modal analysis, the field intensity mode pattern is obtained. Dominant Quasi-TE optical mode is found to be propagating inside the gap regions between nanorods. As discussed in chapter 2, the proportion of power inside the gap regions to the entire cross-sectional area is termed as power confinement. A simplified mathematical equation of confinement factor (Γ) is provided in Eq. 3.1. The optimization of structure is carried out by taking cladding as H_2O without linker and DNA.

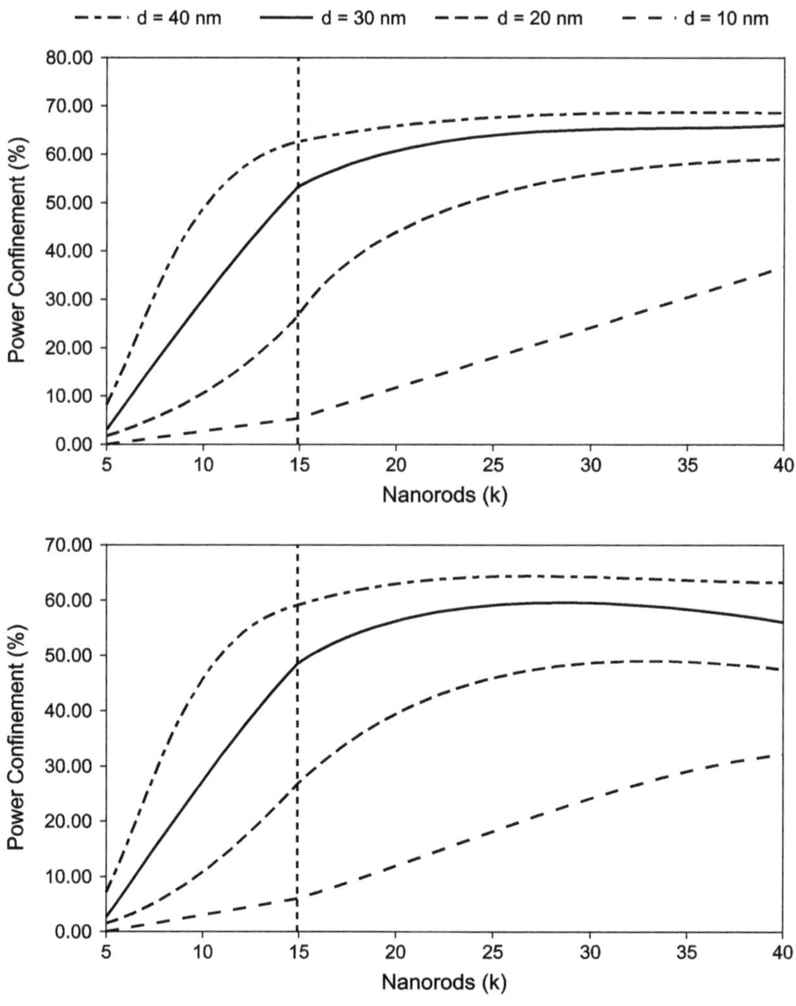

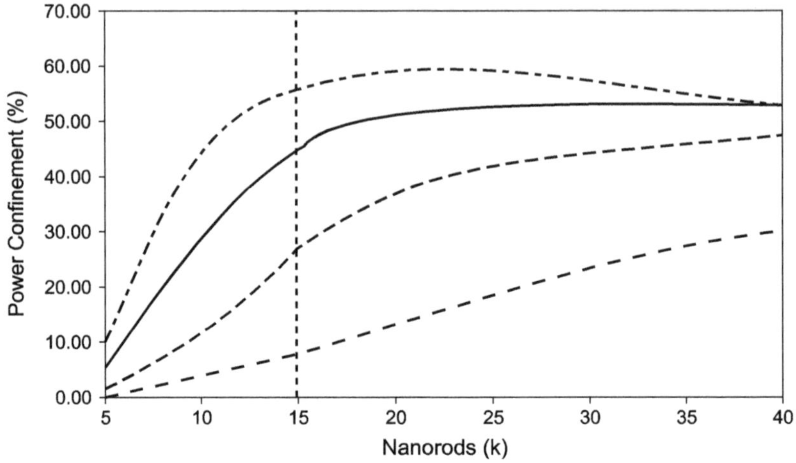

Fig. 8.2. Γ of H_2O filled SNORW cladding with (a) g = 10 nm; (b) g = 15 nm; and (c) g = 20 nm.

The refractive index of Si, SiO_2 and H_2O are 3.4752, 1.444 and 1.31 individually. For illustrative purpose, the height of silicon rods is fixed to 220 nm SOI technology. Certain parameters such as k, g and d are thoroughly examined to provide maximum through-put power. The power confinement is calculated by varying the diameter of silicon nanorods from 10 nm to 40 nm in steps of 10 nm and varying gap between adjacent nanorods from 10 nm to 20 nm in steps of 5 nm. SNORW with k nanowires on SiO_2 substrate is laid in arrayed forest pattern and the k ranging from 5 to 40 is investigated. Power confinement at operating wavelength of 1550 nm for various SNORW parameters combinations are shown in Fig. 8.2. By analyzing Fig. 8.2(a), 8.2(b) and 8.2(c), it can be clearly observed that after $k = 15$ the power confinement gets saturated. Thus, $k = 15$ is most suitable for this structure. Again, by considering g parameter estimation of 10 nm, 15 nm and 20 nm as exhibited in Fig. 8.2(a), 8.2(b) and 8.2(c) respectively, it can be determined that $g = 10$ nm have maximum power confinement inside gap region of SNORW. To satisfy the need of this structure to be compacted, the diameter of each nanowire is calibrated to 30 nm. As it can be seen in Fig. 8.2(a) that maximum power confinement is provided at $d = 40$ nm, however by increasing the diameter to 40 nm, the structure compromises its compactness consequently. So, a better option would be to prefer the diameter = 30 nm to benefit the compactness of sensor. Hence, for the optimization of the SNORW structure for outstanding performance, the parameters k, g and d are calibrated at 15, 10 nm and 30 nm, respectively.

8.4 RESULT AND DISCUSSION

The optimized structure of SNORW with water as cladding material is now implemented for DNA hybridization sensing. Firstly, linker layer of 1 nm thickness on the silicon wire surface is deposited. Subsequently, 8 nm thick layer of DNA is applied over the linker surface. The whole arrangement of SNORW with linker and DNA is thenceforth submerged into water. The refractive index of linker, ssDNA and dsDNA are 1.42, 1.456 and 1.53, respectively (Elhadj, Singh and Saraf, 2004; Steckl et al., 2011). The optical waveguide properties, due to change in refractive index during the hybridization process when double strand DNA splits into single strands of DNA, or two single strands renaturate into a double strand DNA, is thoroughly explored at communication wavelength of 1550 nm, where sources and detectors are readily available.

8.4.1 Mode Field Distribution

The optimized SNORW structure with $k = 15$, $g = 10$ nm and $d = 30$ nm, along with 220 nm standard height of silicon wire with SOI fabrication technology is utilized to get mode field distribution. Field intensity pattern for SNORW is calculated for the dominant fundamental quasi-TE mode and shown in Fig. 8.3(a). Impact on the variation of mode field profile along x-cut axis for 100% ssDNA and 100% dsDNA as sensing measurand layer is shown in Fig. 8.3(b). Optical waveguide to act as an optical bio-sensing device should requisite to vary its mode field intensity pattern by changing the refractive index of measurand layer.

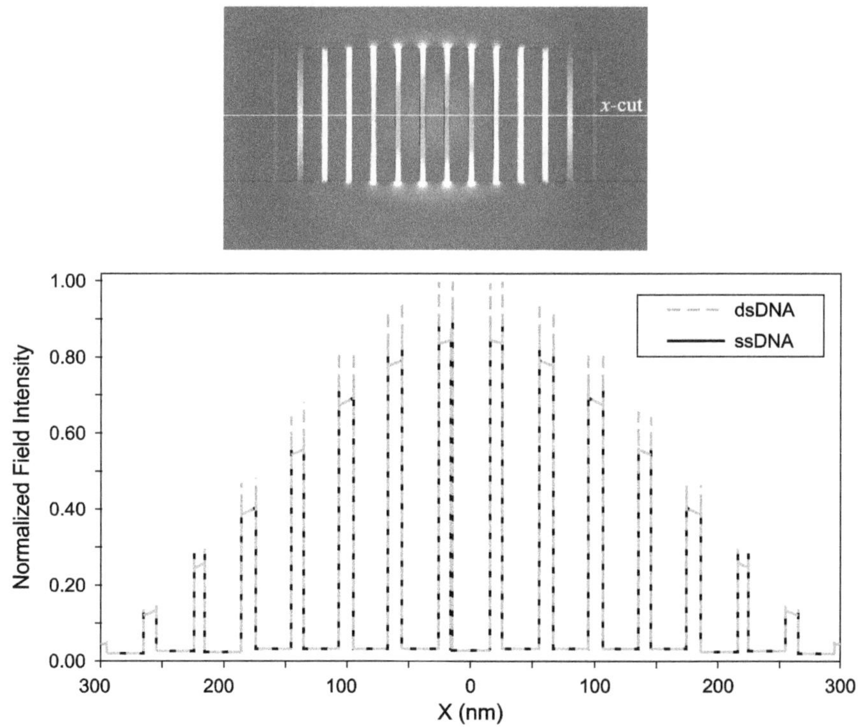

Fig. 8.3. (a) Field Distribution, and (b) *x*-cut field profile of optimized SNORW for 100% ssDNA and dsDNA with $h = 220$ nm, $d = 30$ nm, $g = 10$ nm and $k = 15$.

Field intensity inside linker region is higher than that of DNA region due to comparatively low refractive index of linker than ssDNA and dsDNA, so the spikes are acquired between silicon-wires and DNA layer. Additionally, the level of field intensity in sensing measurand layer increases when dsDNA gets separated into two ssDNA, therefore the intensity inside sensing layer for 100% ssDNA is higher than that of 100% dsDNA. However, increment in the refractive index contrast between silicon nanorods and measurand layer results in decrement of field intensity inside linker layer.

8.4.2 Impact of Waveguide Height

The next parametric variable of SNORW is height of silicon nanorods (h). Height of nanorods are varied from 200 nm to 340 nm while keeping other parameters at its optimum value (i.e. $k = 15$, $d = 30$ nm and $g = 10$ nm). Impact of h on power confinement of SNORW with 100% ssDNA and 100% dsDNA as sensing layer is shown in Fig. 8.4.

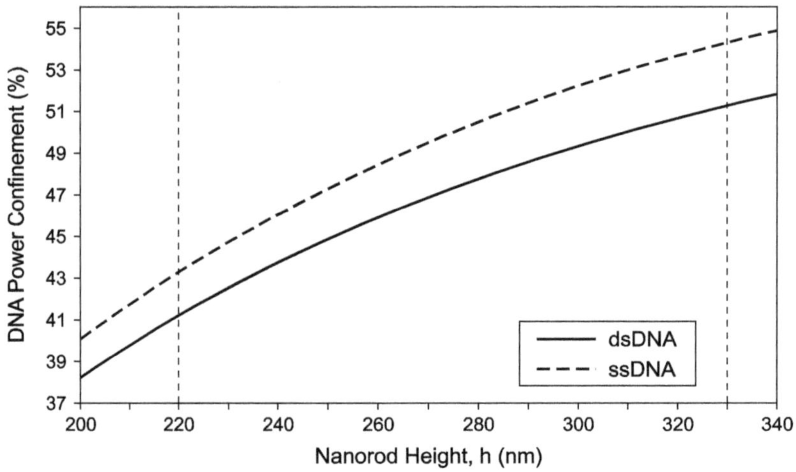

Fig. 8.4. Power confinement in DNA versus h of SNORW with $d = 30$ nm, $g = 10$ nm and $k = 15$.

The increase in h value results in increase in power confinement value. Because of lower refractive index of ssDNA than dsDNA as a sensing layer, former having high power confinement than the latter. For silicon nanorod height ranging from 200 nm to 340 nm, 220 nm and 330 nm are standard silicon height technologies for SOI fabrication. So, both these heights are taken into consideration for better analysis. Power confinement of 43.30% and 54.19% is obtained for ssDNA with $h = 220$ nm and 330 nm, respectively. After complete renaturation to dsDNA, power confinement decreases to 41.22% and 50.64% for $h = 220$ nm and 330 nm, respectively.

8.5 SEPARATION OF dsDNA

The conversion of dsDNA to ssDNA progressively modifies the refractive index of dsDNA-ssDNA concoction. The refractive index for different fraction of dsDNA-ssDNA composition is acquired by Lorentz-Lorenz mixing rule (Tasic *et al.*, 1992) given in Eq. 6.4, which is rearranged to get compositions' refractive index by Eq. 6.5. There, ϕ_a and ϕ_b are volume fractions corresponding to refractive index, n_a and n_b, respectively. Notation '*a*' and '*b*' are used for representing ssDNA and dsDNA, respectively. In integrated optical bio-sensors, any change in chemical parameter of measurand will vary the refractive index. This variation in refractive index affects the effective mode propagation. The probability of detecting various levels of dsDNA-ssDNA in a composition can be estimated by evaluating the guided power confinement in cladding region with change in refractive index due to the mixture. Likewise, the variation in field distribution profile affects the sensitivity

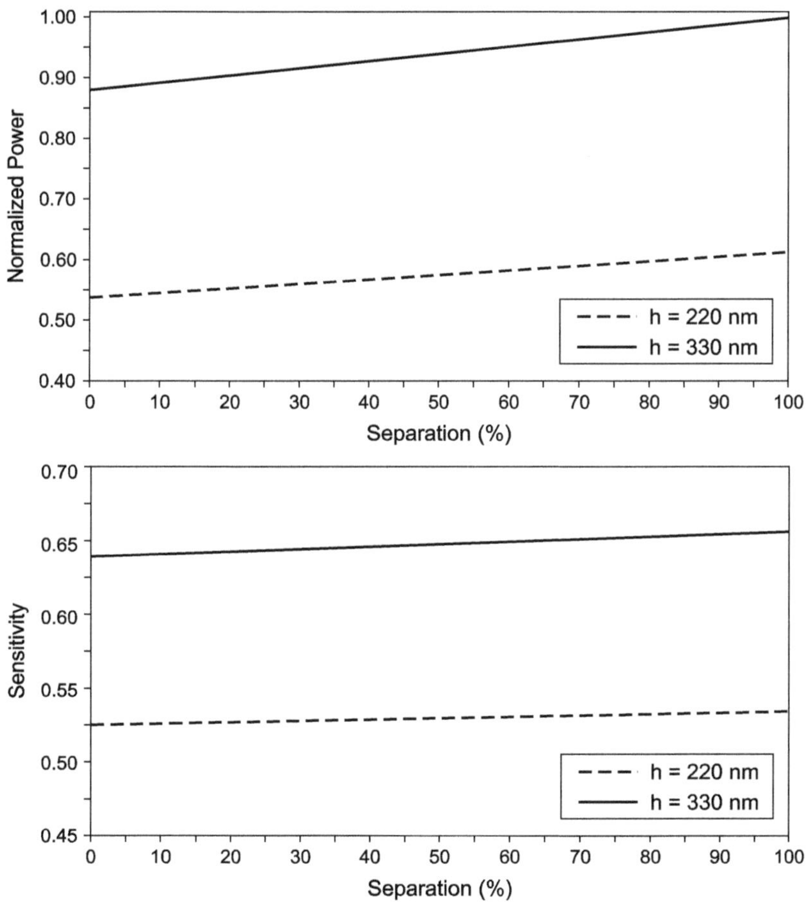

Fig. 8.5. (a) Normalized power, and (b) sensitivity dependence on percentage separation of dsDNA.

performance of the waveguide. So, an optical waveguide to work as a bio-sensor, variation in effective index should vary its guiding power and sensitivity. The normalized power is the ratio of electric field intensity in nanoslot of SNORW for DNA composition to the field intensity in nano-slots for initial DNA type. However, waveguide sensitivity is the ratio of change of effective refractive index of waveguide to the change of refractive index of initial DNA.

Fig. 8.5(a) shows that for SOI standard height technology of $h = 220$ nm, normalized power increases by 3.70% for 50% change in fraction from pure dsDNA to ssDNA-dsDNA composition. The normalized power level significantly increases by 8% when dsDNA completely separates into ssDNA. However, in SOI standard height technology of $h = 330$ nm, 50% change in fraction will lead to increase in normalized power by 5.8%. The normalized power level increases by 12% for complete separation of dsDNA. Waveguide sensitivity corresponding to the separation process is shown in Fig. 8.5(b). By careful observation it can be concluded that SOI technology with 330 nm height is more sensitive towards change in refractive index of DNA sensing layer than that of 220 nm height.

8.6 RENATURATION OF ssDNA

Renaturation is a process in which denatured DNA (ssDNA) can reformulate hydrogen bonds between its complementary strands, making it likely to amend into a form of double helix dsDNA structure. The renaturation of ssDNA to dsDNA will slowly modify the refractive index of ssDNA-dsDNA composition. The probability of detecting various levels of ssDNA-dsDNA composition is estimated by calculating the guided power for effective change in refractive index of cladding region.

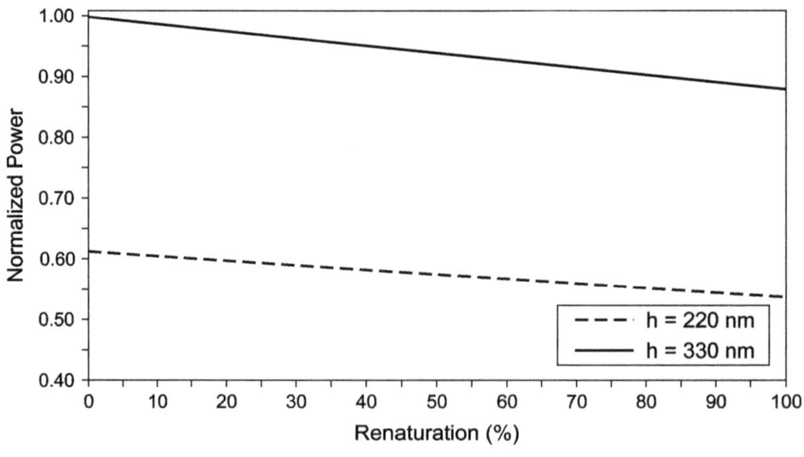

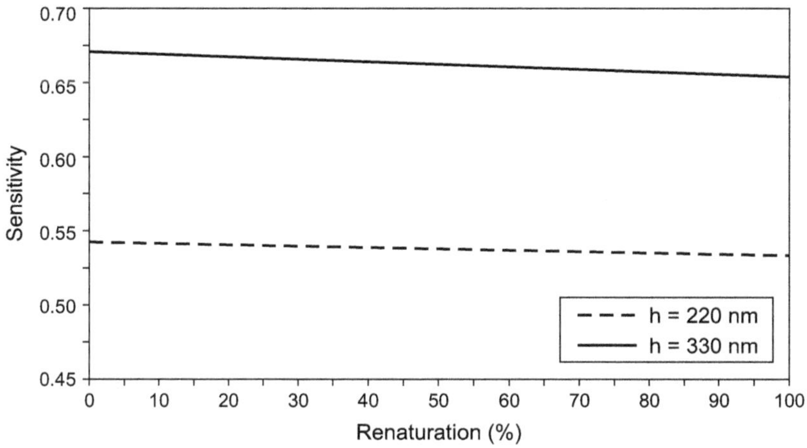

Fig. 8.6. (a) Normalized power, and
(b) sensitivity dependence on percentage renaturation of ssDNA.

Gradual change in fraction from pure ssDNA to dsDNA-ssDNA mixture and its corresponding power levels are shown in Fig. 8.6(a). For SOI standard height technology of $h = 220$ nm, the normalized power level decreases by 8% when ssDNA completely renature to dsDNA. While, in SOI standard height technology of $h = 330$ nm, the normalized power level decreases by 12% for complete renaturation. Sensitivity of SNORW towards renaturation is shown in Fig. 8.6(b). The waveguide sensitivity in renaturation process decreases with the increase in percentage of renaturation. However, in case of renaturation, 220 nm height technology is more sensitive than that of 330 nm height technology. Precise and fast detection of DNA hybridization are very crucial for which SNORW will be extremely compelling solution.

8.7 SUMMARY OF DNA HYBRIDIZATION SENSING

Silicon nanowire optical rectangular waveguide based biosensor with advanced characteristics has been suggested to detect DNA hybridization. Characteristic properties like power confinement, mode field intensity and normalized power is been studied by applying full vectorial finite element method. The optimum power confinement is been obtained by appropriately designing the structure parameters inside water as a cladding material. For high power confinement and physical compactness, $k = 15$, $d = 30$ nm and $g = 10$ nm were obtained as optimized SNORW parameters.

Mode field intensity is being obtained by adding linker and DNA layer to SNORW structure, which shows that any change in refractive index is detected with corresponding change in field intensity level. Silicon wires of height 220 nm and 330 nm is been considered corresponding to standard silicon fabrication technology used for DNA hybridization. The change in normalized power is ascertained in case of separation of dsDNA strands, when double strand DNA splits into two single strand DNA. Also, the case of renaturation, when two single strand DNA combines to form a double strand DNA, is been thoroughly investigated with a fractional change in the composition of ssDNA and dsDNA. The difference in the power confinement in presence of 100% ssDNA and 100% dsDNA is being found as 8% for 220 nm and 12% for 330 nm SOI technology, respectively. Renaturation and separation of DNAs are slow processes, in which the power confinement is been evaluated and compared, when different percentages of dsDNA and ssDNA coexisted.

❑❑❑

9 SoI Ring Resonator for Optical Filter

Silicon photonics due to its remarkable progress and exemplary performance in various applications such as on-chip communication and sensing application, is attracting various researchers in this field. Generally, standard integrated waveguide height of 220 nm is used in silicon photonics category of PIC for best optimized connectivity between two on-chip devices. In order to employ this technology, it is required to further optimize the degree of freedom in other dimensions of waveguide. SOI having high index contrast, which is compact and compatible to CMOS fabrication technology is being used to construct ring-based structure for different applications, i.e. optical filter for on-chip communication. These structures are known as ring resonator can be used as multiplexing or de-multiplexing unit according to their resonance wavelength criteria and structural arrangement.

The aim of this chapter is to study several applications of optical ring resonator. Its applications as a notch filter is broadly analysed. The different design parameters of an optical ring resonators such as Free Spectral Range (FSR), Full Width at Half Maximum (FWHM), finesse and quality-factor, are explored in detail, which plays an imperative role in different applications.

The basic theory of optical ring resonator is studied and the operation of SOI ring resonator as a notch filter is analyzed. The designing is done in such a way, as to have an ultrafast notch filter. The ring resonator's performance parameters such as FSR, FWHM, finesse and quality-factor are investigated to acquire the required characteristics of the optical notch filter. Numerical simulations using FEM and its corresponding mathematical analysis are performed in order to design an ultrafast optical ring resonator notch filter.

9.1 INTRODUCTION TO OPTICAL RING RESONATOR

A growing demand in today's optical telecommunication is, innovation to proceed to deliver integrated photonic devices that are cost efficient, small in size and have higher bandwidth. Optical Ring Resonators have attracted the great interest in advancement in optical communication to offer a variety of application such as filters, multiplexers/de-multiplexers, optical routers, optical signal processor, etc. (*Integrated Ring Resonators*, 2007; Van, 2017). Recently, optical ring resonators

DOI: 10.1201/9781003783589-9

have also found their way to be used as sensors and biosensors (Khorasaninejad and Saini, 2011). Integrated ring resonators do not require facets or gratings for optical feedback and are thus particularly suited for monolithic integration with other components. The response from coupled ring resonators can be custom designed by the use of different coupling configurations. In this way, the response from ring resonator filters can be designed to have a flat top and steep roll of. Other coupling configurations lead to a ring resonator reflector device, which can be used on one side or on both sides of an integrated semiconductor optical amplifier serving as a laser.

An optical ring resonator is a set of waveguides in which at least one is a closed loop coupled to some sort of light input and output. Ring resonators play an important role in the success of silicon photonics, because silicon enables ring resonators of an unprecedented small size. A generic ring resonator consists of an optical waveguide which is looped back on itself, such that a resonance occurs when the optical path length of the resonator is exactly a whole number of wavelengths. One of the important features of ring resonator is the resonator frequency as it can operate only at a particular frequency. So this is a frequency dependent device which means that it is beneficial to a lot of applications.

9.2 THEORY OF OPTICAL RING RESONATOR

In general, a ring resonator consists of a looped optical waveguide and a coupling mechanism to access the loop. When the waves in the loop build up a round trip phase shift that equals an integer times 2π, the waves interfere constructively and the cavity is in resonance. The coupling takes place only if the wavelength satisfies the resonance condition (Wageeh, El-Sabban and Khalil, 1995).

$$2\pi R n_{eff} = m\lambda_m \quad ...(9.1)$$

Where R is ring radius, n_{eff} is effective refractive index, m is an integer representing the mode order, and λ_m is the resonance wavelength of mode 'm'.

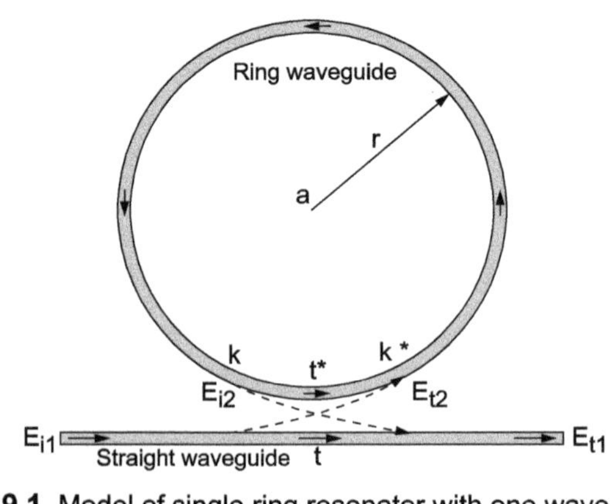

Fig. 9.1. Model of single ring resonator with one waveguide.

There are three aspects that affect optical coupling they are: (a) distance, (b) coupling length, (c) refractive indices between waveguide and the optical ring resonator. In order to optimize this coupling, it is usually a case to narrow the distance between ring resonator and waveguide. Thus, closer the distance, the easier the optical coupling happens. In addition, the coupling length affects coupling as well. The coupling length represents the effective curve length of ring resonator for the coupling phenomenon to happen with waveguide. It has been studied that as the optical coupling length increases, the difficulty for coupling to happen decreases (Marcatili, 1969; Halir *et al.*, 2012; Yun *et al.*, 2016). Furthermore, the refractive index of waveguide material, ring resonator material and medium material in between waveguide and ring resonator also affect the optical coupling. The medium material is usually an important one been studied since it has a great effect on the transmission of the light wave. The refractive index of the medium can be either large or small according to various applications and purposes.

Fig. 9.1 shows the basic configuration of a ring resonator. It consists of ring waveguide of radius '*r*' and a straight waveguide. Coupling occurs when the resonance condition is satisfied. When a single unidirectional mode of the resonator is excited then coupling is lossless and single polarization is considered as the coupling of waves with different polarization is not transpired by coupling of waveguide segments and coupler elements. The various kinds of losses occurring along the propagation of light in ring resonator filter are incorporated in attenuation constant, the interaction can be described by matrix relation:

$$\begin{pmatrix} E_{t_1} \\ E_{t_2} \end{pmatrix} = \begin{pmatrix} t & k \\ -k^* & t^* \end{pmatrix} \begin{pmatrix} E_{i_1} \\ E_{i_2} \end{pmatrix} \quad \text{...(9.2)}$$

Where, E is the field intensity with E_{tn} and E_{in} denoting transmitted and input field for $n = 1$ and $n = 2$, which are for straight and ring waveguide, respectively. The coupler parameters 't' and 'k' denote the value of transmittance and coupling coefficient. The '*' denotes the conjugated complex value of 't' and 'k', respectively. As shown in Fig. 9.1, 'a' is the loss coefficient of the ring, which is considered as 1 for zero loss.

Ring resonators can be described by certain design parameters on which the functionality of the device depends. The important design parameters are: Free Spectral Range (FSR), Full Width at Half Maxima (FWHM), Finesse and Quality Factor. FSR is one of the key specifications of Ring Resonator. It is mainly defined as the distance between two resonance peaks (*Integrated Ring Resonators*, 2007):

$$\text{FSR} = \frac{\lambda_0^2}{L.n_{eff}} = \frac{\lambda_0^2}{2\pi R.n_{eff}} \quad \text{...(9.3)}$$

Where, λ_0 is operating wavelength and L is circumference of the ring. FWHM is the 3dB bandwidth of resonance wavelength spectra with line width of half the amplitude of resonance wavelength. The expression for FWHM can be given as (*Integrated Ring Resonators*, 2007);

$$\text{FWHM} = \frac{K^2\lambda_0^2}{\pi L n_{eff}} \quad ...(9.4)$$

Where, K is the coupling coefficient. Finesse (F) is a measure of the resonator's frequency selectivity. It is defined as the ratio of filter's FSR to FWHM, the width of a resonance for a specific wavelength (*Integrated Ring Resonators*, 2007).

$$F = \frac{FSR}{FWHM} \quad ...(9.5)$$

Quality Factor (Q) of a resonator is a measure of sharpness of the resonance. It is defined as:

$$Q = \frac{L.n_{eff}}{\lambda_0} F \quad ...(9.6)$$

9.3 RECTANGULAR WAVEGUIDE RING RESONATOR

The basic configuration of ring resonator which consists of unidirectional coupling between ring and a straight waveguide is typically referred to as a Notch Filter. In this section, both the straight bus waveguide and ring waveguide are considered as ridge waveguide. The similar SOI case as of previous chapters having 220 nm of waveguide height is considered in this chapter. Here, the resonance wavelength is only allowed to couple inside the ring. While rest of the wavelengths are passed through the straight waveguide and obtained at the through port.

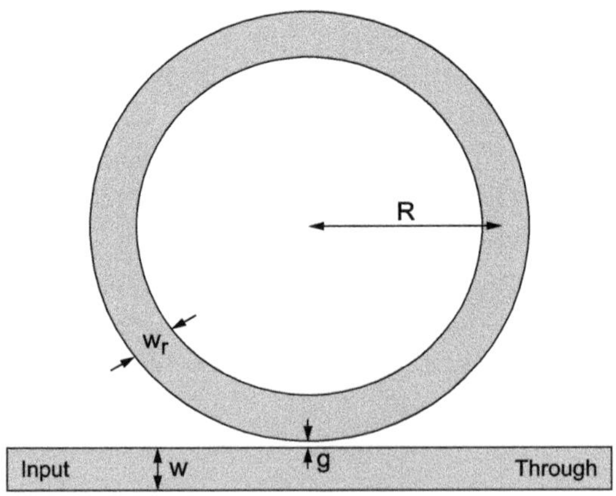

Fig. 9.2. Structure of a conventional ring resonator waveguide.

The basic configuration of ring resonator is taken which consists of unidirectional coupling between ring resonator and a straight waveguide. A simple configuration is obtained using SOI technology by taking an advantage of high effective index difference between Si and SiO_2 which provides natural vertical optical confinement. In this design, R is defined as the ring radius from the center of ring to the center of waveguide. The width of ring waveguide (W_r) is 400 nm and of bus waveguide (W) is 400 nm. Separation (g) between the bent waveguide and straight waveguide is adjusted to 100 nm as shown in Fig. 9.2. The refractive indices of Si and SiO_2 are taken as $n_{Si} = 3.4752$ and $n_{SiO2} = 1.444$, respectively. Where, the variation due to dispersion is ignored. The analysis is being done for TE mode by providing 10 dBm power at the input port of straight waveguide.

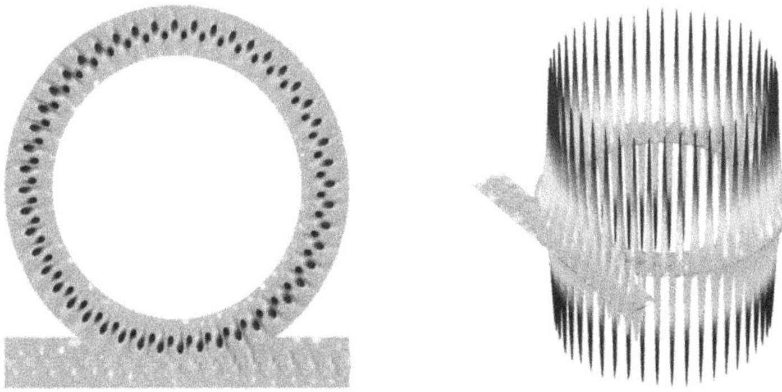

Fig. 9.3. E-field confinement in ring resonator waveguide at the resonance wavelength (a) top view; and (b) 3-D view.

Fig. 9.3 represents the condition of resonance that is satisfied for 1550 nm of wavelength, thus for this resonance wavelength, all the power is confined inside the ring waveguide itself. The illustrative waveguide parameters used to obtain this 1550 nm wavelength as resonance wavelength are: ridge waveguide width of 400 nm and ring radius of 20 μm. Due to this ridge waveguide geometry, the optical intensity is confined in the high refractive indexed silicon region. The fused silica is used to cover the surrounding empty space, which act as insulating layer and protect the effect of nearby environment on guiding properties of waveguide. The detailed analysis of parametric properties of ring resonator with its dependency on ring radius is performed in the subsequent sections.

9.4 SNORW RING RESONATOR

The configuration of ring resonator is discussed in this section which consists vertical silicon nanowires only. In this structure bus waveguide and ring waveguide both are designed using silicon nanowire array. For better understanding and easy comparison of this SNORW based ring resonator with the above mentioned rectangular waveguide based ring resonator, the overall width of the waveguide is kept same (i.e. 400 nm). SNORW configuration is obtained using SOI technology by taking an advantage of high effective index difference between Si and SiO_2 which provides natural vertical optical confinement in the gaps of silicon nanowires.

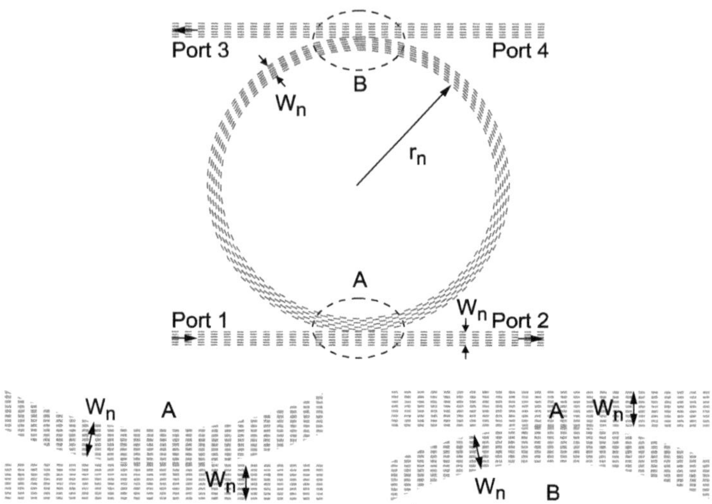

Fig. 9.4. (a) Structure of a SNORW based ring resonator, (b) zoom-in view of encircle-A, and (c) zoom-in view of encircle-B.

In this design, r_n is defined as the ring radius from the center of ring to the center of SNORW ring waveguide. The width of SNORW (W_n) is 400 nm as shown in Fig. 9.4(a). Separation between the ring waveguide and straight waveguide is adjusted to 0 nm, so that maximum coupling effect will be achieved between straight bus and ring. A clear view of lower contact between bus waveguide and ring waveguide can be seen in Fig. 9.4(b) and (c). In SNORW, a group of 7 silicon nanowires are taken in an array of waveguide with pitch size of 1.5 times nanowire diameter. The nanowire diameter is taken as 40 nm, which result in 60 nm pitch size. The pitch size

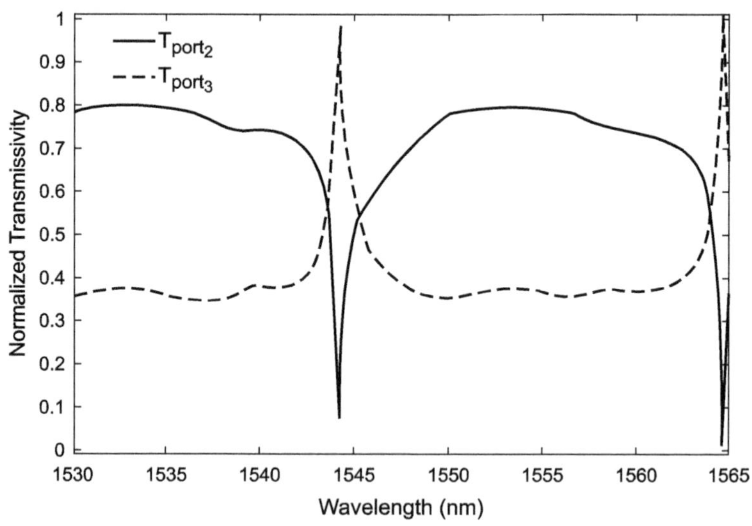

Fig. 9.5. Wavelength versus normalized power transmissivity at port 2 and 3.

is the distance between two nanowires from center of each wire, which means that the gapping between boundary to boundary of silicon nanowire is of 20 nm. Thus, the whole configuration of this waveguide will result in total waveguide width of 400 nm. The value of r_n is taken as 5 μm, thus the distance between two waveguides opposite to each other is 5.4 μm. In the construction of Fig. 9.4(a), 210 nanowires are used in designing of each straight bus waveguide and 3535 number of nanowires are used in designing of ring portion, which means that total of 3955 nanowires are used in designing of whole structure. The finite element method (FEM) is used for simulating this structure. The minimum mesh element size of 2 nm is considered in the silicon nanowire and gapping region, which is 1/20 times diameter of nanowire. In the same region, the maximum mesh element size is taken as 4nm, which is 1/10 times nanowire diameter. The maximum mesh element growth rate and curvature factor are taken as 1.1 and 0.2, respectively. Solutions of electromagnetic waves in frequency domain is solved for 10449591 number of triangular mesh elements.

Sellmeier equation for refractive index of silicon (n_{Si}) from Eqn. 3.9 and fused silica (n_{SiO_2}) Eqn. 3.10 are used in this simulation, so to consider wavelength dependency on refractive index. Wavelength sweep is applied at the input port-1 from 1530 nm to 1565 nm to cover the range of optical C-band wavelengths. Resolution of 0.1 nm is applied during this wavelength sweep to get a proper vision of coupled spectrum. The spectrum for wavelength transmitted at output port (port-2) of bus waveguide and port-3 of filtered waveguide is shown in Fig. 9.5 with red and blue colour, respectively. This spectrum clearly indicates that the resonance wavelength of this SNORW based ring resonator is 1544 nm. The computational time and memory used for each wavelength step of 0.1 nm are 5 minutes and 29 GB, respectively. Thus, it can be easy understood that the complexity for solving this structure is

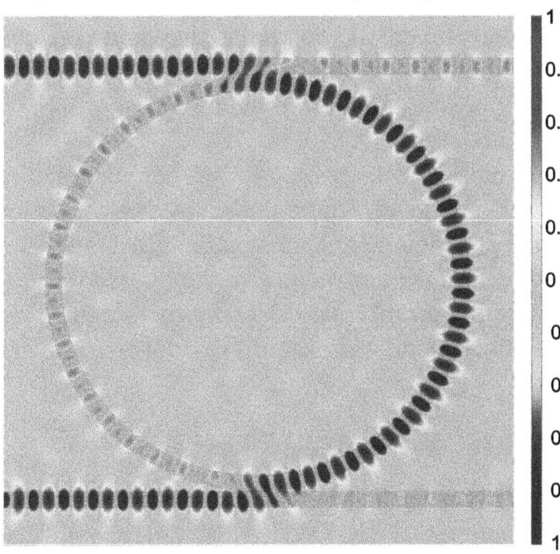

Fig. 9.6. Normalized E-field in SNORW based ring resonator at the resonance wavelength.

very high, which takes total computational runtime of 29 hours and 15 minutes. In Fig. 9.5, the second resonance wavelength in the spectrum is 1564.5 nm, this means that the FSR of 20.5 nm is achieved in this SNORW ring resonator.

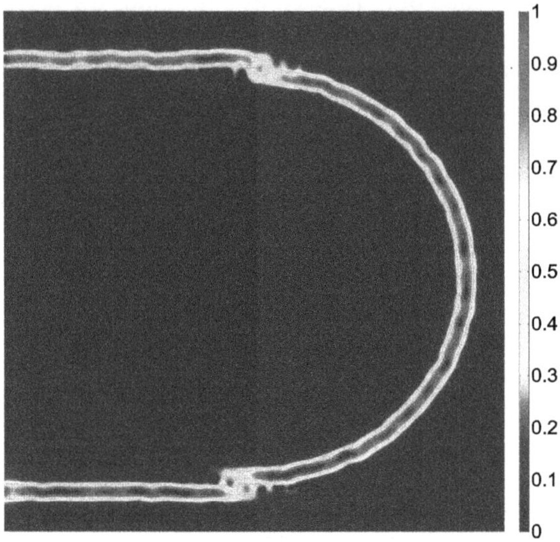

Fig. 9.7. Normalized power confinement in SNORW ring resonator at the resonance wavelength.

An identical arrangement of silicon nanowires in bus and ring waveguide makes the mode propagation constant same for both bus and ring waveguide. The normalized electric field distribution for the top cross-section view is shown in Fig. 9.6. This normalized E-field is shown for the propagation of 1544 nm wavelength from port-1 to port-3 of SNORW ring resonator. The normalized power flow in the structure at resonance wavelength of 1544 nm is shown in Fig. 9.7. The input wave travels half way through the bus waveguide, curve over the ring and exit through the output waveguide, in opposite direction to the direction of input wave, whenever resonance condition is acquired.

Replacing silicon nanowire technology with conventional integrated optics takes the photonics to a new level, where the cavity between nanowires serves as a nice optical confinement region. The advantage of using silicon nanowire over ring resonator can be seen in the narrowness of the filtered wavelength ranges and hence the sharpness of the filtering process. This structure finds its application as a notch filter to filter out the C-band wavelengths by adjusting waveguide geometrical parameters.

9.5 PERFORMANCE PARAMETERS

The purpose of this work is to model a SOI ring resonator notch filter and investigate its important design parameters (FSR, FWHM, F and Q) to optimize its performance by achieving the appropriate values of characteristics such as; Minimum value of

FWHM, high value of finesse and maximum value of Quality factor. Numerical simulations using Finite Element Method (FEM) and mathematical analysis is performed in order to design an ultrafast optical ring resonator notch filter. C-band (1530 nm to 1565 nm) is taken as the range of wavelength to attain the desired values of the filter characteristics of ring resonator for its enhanced performance. For this SOI ring resonator structure, the values of propagation constants (β) have been extracted using FEM for a range of ring radius from 7 μm - 40 μm. Pointing vector from FEM is obtained for c-band wavelength and the exact positions of the peaks corresponding to the wavelengths is calculated in order to have an understanding about impact of varying radius on FSR and FWHM.

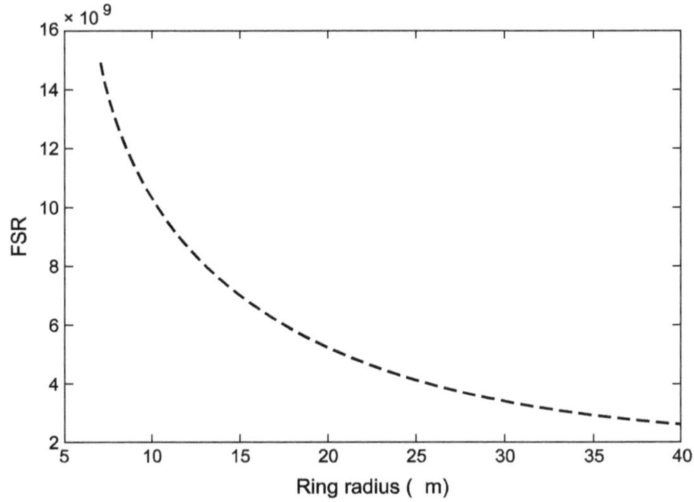

Fig. 9.8. FSR vs ring radius.

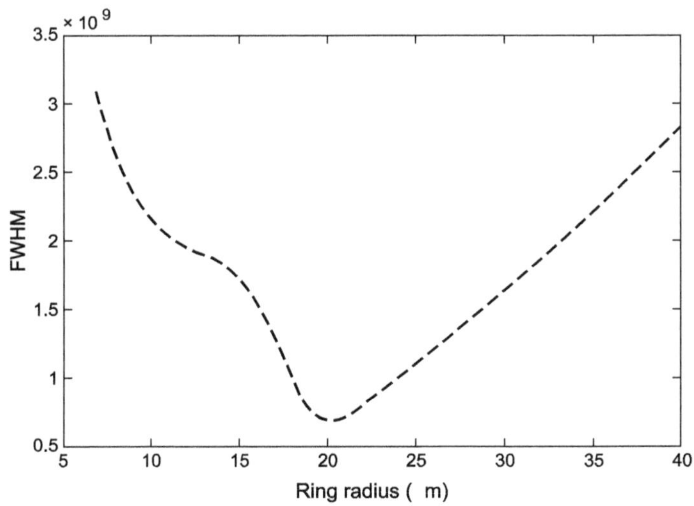

Fig. 9.9. FWHM vs ring radius

Fig. 9.8 shows the variation in FSR as the ring radius changes. It can be clearly observed from the plot that FSR is inversely proportional to the ring radius. The calculated value of FSR for R = 20 μm is 5.2 nm.

Fig. 9.9 represents that the profile of FWHM has its minima at 20 μm of ring radius, which is always required for optimization of any structure. The calculated minimum value of FWHM for this structure is 0.6844 nm.

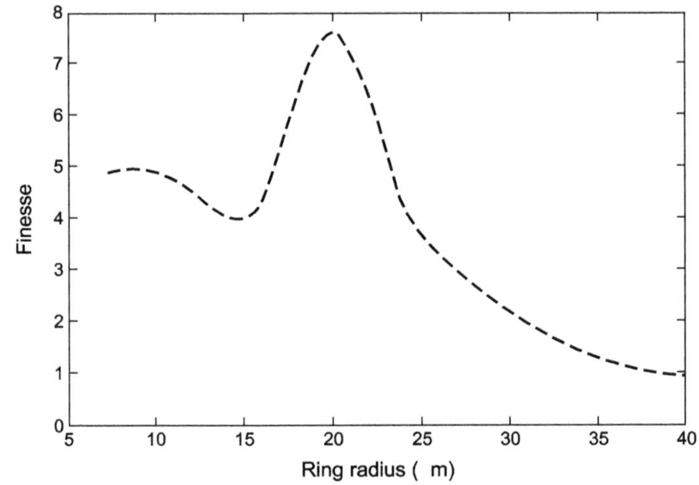

Fig. 9.10. Finesse vs ring radius.

From Fig. 9.10, it can be observed that at 20 μm ring radius, the value of finesse is maximum. The calculated maximum value of Finesse is F =7.5975. High value of F corresponds to narrow peaks.

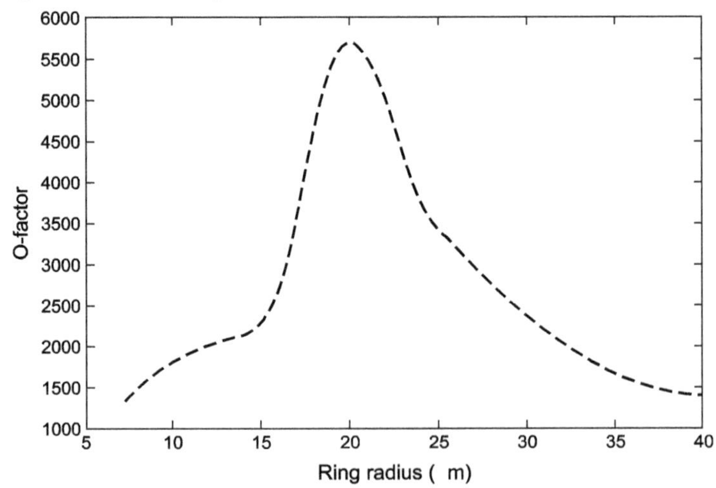

Fig. 9.11. Quality Factor vs ring radius.

Fig. 9.11 shows that the Quality factor is also maximum for 20 μm ring radius. Quality Factor of a resonator is a measure of the sharpness at the resonance.

The calculated value is Q = 5692. There is no variation in the propagation constant (β) as ring radius changes. n_{eff} will also be unaffected as β remains unchanged.

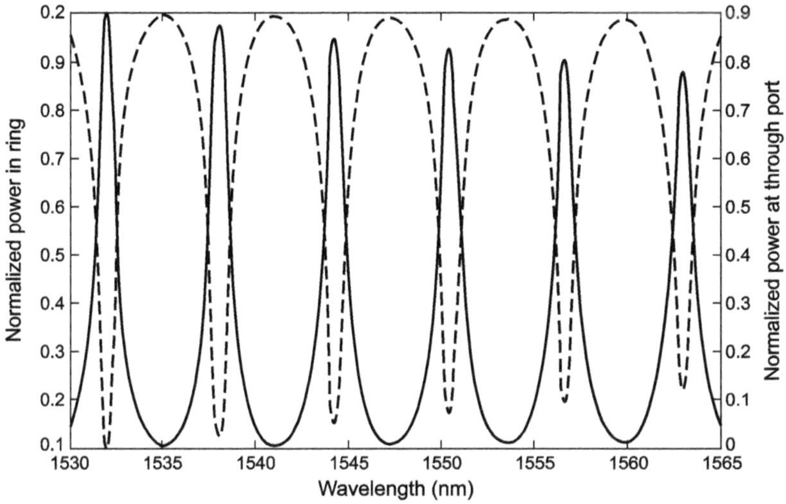

Fig. 9.12. Wavelength versus normalized power in ring and at through port.

One of the important observations with the help of Fig. 9.12 is the occurrence of six peaks for 20 μm radius. The wavelength corresponding to these peaks are resonance wavelengths for this range. For C-band, resonance wavelengths λ_1, λ_2, λ_3, λ_4, λ_5 and λ_6 are 1535 nm, 1540.2 nm, 1545.5 nm, 1550.7 nm, 1556 nm and 1561.4 nm, respectively. The resonance wavelengths are the one, which can be coupled in the ring waveguide. All wavelengths except resonance wavelength are passed through the output of straight waveguide. 1550 nm of wavelength can also be filtered, which has its several applications in the field of optical communication.

9.6 SUMMARY OF SOI RING RESONATORS

The operation of Silicon-on-Insulator (SOI) ring resonator as a notch filter is analysed. The designing is done in such a way as to have an ultrafast notch filter. The ring resonator's performance parameters such as Free Spectral Range (FSR), Full Width at Half Maximum (FWHM), finesse and quality-factor are investigated to get the optimized values of the characteristics of filter. Considering the range of ring radius, corresponding optimum values of FSR, FWHM, Finesse and Quality factors are investigated, which can encompass the desired performance of filter characteristics. The characteristics and performance parameters of single bus waveguide notch filter are studied. Simulations and calculations are carried out using FEM and analytical equations. This can be concluded that for a range of ring radius taken as 7μm – 40μm, 20μm radius has minimum value of FWHM, maximum value of Finesse, and a high value of Quality factor. The designed structure succeeded in optimizing the performance parameters.

10 Ring Resonator Application for Glucose Sensing

In this chapter, a silicon-on-insulator based ring resonator is used for sensing application. An optical ring resonator based glucose sensor, which is found to be miniaturized, highly sensitive, precise and reliable. This sensor is for spot glucose level monitoring in blood over the range of 10 mg/dl – 200 mg/dl using minimal invasive approach. Two different schematics of this ring resonator sensor is been designed based on the fundamentals of rectangular waveguide structure, which is further analysed and discussed. First schematic is based on ridge waveguide in the ring while other is based on slot waveguide. Their sensing performance is then compared with the sensor consisting of ridge and slot waveguide, respectively. The slot waveguide based sensor can predict more accurate result leading to a future development in optical glucose sensing methods. The changing concentration of glucose affects the refractive index of blood sample, which results in a wavelength shift exhibiting deviation in response of the sensor.

10.1 ADVANCEMENT IN OPTICAL BIOSENSORS

In the past few decades, optical biosensors, for their promising applications have gained a significant amount of attention in medical diagnostics (Clinical Monitoring), agro-industry, environmental and water monitoring and food processing industry (Marazuela and Moreno-Bondi, 2002; Mehrotra, 2016). Comprising of compact, highly accurate, reliable, fast and highly sensitive characteristics, optical biosensor serves its purpose. Based on their detection techniques, biosensors can be mainly classified as label-based and label-free biosensors (Cooper, 2009). Label-free biosensor is more advantageous than label-based biosensor because of no requirement of labelling and complex handling of liquid (Dar *et al.*, 2012). So, label free optical methods for the detection of bio-chemical analytes is anticipated to have a great future in real time domain with less assay time requirement.

A lot of research investigations have been performed on biosensors incorporated with different waveguides. An optical biochemical sensor using silicon multiple slot waveguide is analysed in this section and some of the parameters such as ridge width, slot width, the number of slots etc. have been optimized to achieve the highest possible sensitivity (Khodadad *et al.*, 2014). Further, this multiple-slot structure is

DOI: 10.1201/9781003783589-10

studied in a ring resonator sensor. A horizontal slot waveguide is been designed as a biosensor for detection of DNA hybridization (Viphavakit *et al.*, 2015). A SOI ring resonator biosensor that uses sub wavelength grating enhances the sensitivity and detection limit (Flueckiger *et al.*, 2016). A compact sagnac interferometer based cost effective biosensor for glucose detection is demonstrated with a high average sensitivity (An *et al.*, 2017). In recent years, silicon-on-insulator (SOI) based sensors have gained tremendous potential because they provide the way to realize a commercially feasible, miniaturized and highly sensitive photonic sensor.

Diabetes mellitus, a non-curable disease has many people suffering from it. Having a precise glucose sensor for spot glucose monitoring will significantly make routine glucose measurement more convenient. Using the optical method for glucose detection would offer fast and frequent measurement and accurate sensing. Minimal invasive and Non-invasive methods (Koschinsky and Heinemann, 2001) are the two approaches mainly adopted. Minimal invasive method includes transdermal approaches, glucose electrodes, micro dialysis and open flow micro perfusion. Some of the glucose monitoring devices available in today's date is found to have used Transdermal approach. For example: Gluco-watch biographer, in which a low current is applied to the skin which drags ions together with interstitial fluid through the skin. However, the applied current causes local skin irritation which can last 3-27 days, and the device fails to work in case of increased sweating. SpectRx (Narcross, GS, USA) developed a system that works on the principle that it continuously drags interstitial fluid over some days by means of a laser beam. The main issue with this system is time lag (Tamada, Bohannon and Potts, 1995). Glucose electrodes were designed to be inserted into the skin for the first time, under minimal invasive approach (Gough and Armour, 1995). The basic principle of this method involves immobilization of glucose oxidase on the tip of a needle, which is made up of an insulating material sandwiched between two different metals. Then glucose is monitored by measuring the change in flow of current in interstitial fluid. But the disadvantage of this technique is its long-term instability which is due to lack of biocompatibility of electrode surface. Another method is micro-dialysis (Summers *et al.*, 1999; Koschinsky and Heinemann, 2001). In this method a semi–permeable dialysis Fibre is inserted into the subcutaneous tissues and then this fibre is perfused with isotonic fluid. Dialysate of the interstitial fluid is being collected and pumped to a glucose sensor present outside of the body for the detection of glucose level. But the major disadvantage of this method is its physical time lag which varies between 5 to 45 minutes. The non-invasive approach mainly uses spectroscopic techniques (Khalil, 1999; Koschinsky and Heinemann, 2001) and scattering (Alarousu *et al.*, 2004). A typical absorption spectrum can be obtained by directing a light beam through a solution of glucose in water. Large number of works have been offered to use spectrophotometer approach for measuring glucose in human skin. This approach is based on solute specific absorption bands in different spectral range to quantify solutes in liquids. However, it is challenging to achieve precision for a range of blood glucose for a prolonged period of time because changes in blood glucose can induce a very small and unspecific change in the absorption spectra. On the other hand, scattering is one of the drawbacks of spectroscopic approach because human skin has different structures. Biosensors, by means of photonic integrated on-chip circuits, have been offered in recent trends. A widely adopted approach includes biosensors based on rectangular waveguide and slot waveguide (Li *et al.*, 2017). The rectangular waveguides where light is confined in the high refractive indexed region are applicable as biosensors due to their high refractive index contrast and less transmission loss (Mukundan *et al.*, 2009; Kozma *et al.*, 2014). But, these biosensors have limited sensitivity because of weaker overlap between light and sensing material. In contrast to this, adopting the slot waveguide would improve the sensitivity as it has the light, strongly confined, and guided in low refractive index area, which would result into the light field distribution which is rather sensitive to the refractive index of the measurand.

10.2 METHODOLOGY IN SENSING GLUCOSE

In this work, silicon-on-insulator technique provides a platform for photonic integrated circuit sensing based on minimal invasive approach. For serving this purpose, two of the waveguides, ridge waveguide and slot waveguide based optical ring resonators are taken into consideration for comparison of sensing characteristics. The aim is to perform glucose sensing by taking only 0.1 fraction of blood sample as measurand. Entire optimization is performed using Finite Element Method (FEM) parametric analysis. Afterwards, the key parameters of the ring resonator like Free Spectral Range (FSR), Full Width at Half Maximum (FWHM), Finesse (F), Quality-factor (Q-factor) of coupled and filtered spectrum is calculated. The important characteristics of this glucose sensor which have been observed here are sensitivity, confinement factor, and wavelength shift. All these characteristics have been evaluated with respect to changing glucose concentration for both ridge and slot waveguide based ring resonators and then compared accordingly.

For the detection purpose, primarily, the evaluation is done for an empty waveguide case i.e. ring cavity material is taken as air and further this ring cavity is replaced by the sample. A measurement sample is taken with a very little amount of blood glucose as measurand, taken in the fraction of 0.1. It is then mixed with 0.8 fraction of Ethylenediaminetetraacetic acid (EDTA). EDTA and many other compounds act as an anticoagulant agent such as sodium acetate, heparin, potassium oxalate etc. The refractive index of EDTA at 1550 nm is given as 1.363. Finally, 0.1 fraction of sodium fluoride (NaF) is added as a preservative in the sample. The calculated refractive index value of NaF is 1.3194 (Li, 1976) at 1550 nm. The refractive index of blood (Yahya and Saghir, 2016; Lazareva and Tuchin, 2018) has been evaluated using

$$n_{\text{blood}}(\lambda) = \sqrt{1 + \frac{A_1\lambda^2}{\lambda^2 - B_1} + \frac{A_2\lambda^2}{\lambda^2 - B_2}} \qquad \text{...(10.1)}$$

Here, the value of empirical constants A_1, A_2, B_1, B_2 are 0.83423, 0.04296, 10775.44775 and 6.13587E6, respectively. Using Eq.1, the value of blood refractive index at 1550 nm is 1.3456. The average refractive index of glucose solution with different concentration of glucose (Yeh, 2008) has been evaluated using

$$n_{gs} = (0.00011889 \times c) + 1.33230545 \qquad \text{...(10.2)}$$

Where c is glucose concentration (mg/dl) and n_{gs} is the average refractive index of glucose solution. By assuming 0.2 and 0.8 fraction of glucose and blood, using Gladstone-Dale mixing rule, new refractive index of blood sample having different glucose concentration has been calculated as:

$$n_{bg} = 0.2n_{gs} + 0.8n_{\text{blood}} \qquad \text{...(10.3)}$$

Here, n_{bg}, n_{gs}, n_{blood} are the refractive index of the blood glucose, glucose solution with different concentration, and blood, respectively. Since we are mixing three components (blood glucose, EDTA, NaF) together, thus to evaluate the refractive index of the mixture, an appropriate mixing rule is required. For this, we have taken Lorentz-Lorenz mixing rule (Tasic *et al.*, 1992) into consideration. This rule is frequently used for refractive index analysis because of its accuracy and refractive index value which is closer to practical data and has only a small deviation from the experimental value. Lorentz-Lorenz equation for a mixture comprising of three components is given as:

$$\frac{n_m^2 - 1}{n_m^2 + 2} = \phi_1\left(\frac{n_1^2 - 1}{n_1^2 + 2}\right) + \phi_2\left(\frac{n_2^2 - 1}{n_2^2 + 2}\right) + \phi_3\left(\frac{n_3^2 - 1}{n_3^2 + 2}\right) \qquad \text{...(10.4)}$$

$$n_m = \sqrt{\frac{1 + 2A + 2B + 2C}{1 - A - B - C}}$$

$$A = \phi_1\left(\frac{n_1^2 - 1}{n_1^2 + 2}\right),\quad B = \phi_2\left(\frac{n_2^2 - 1}{n_2^2 + 2}\right) \text{ and } C = \phi_3\left(\frac{n_3^2 - 1}{n_3^2 + 2}\right)$$

Here, n_m is the resultant refractive index of the mixture, n_1, n_2 and n_3 are the refractive indices of NaF, EDTA and blood glucose, and ϕ_1, ϕ_2 and ϕ_3 are volume fraction of NaF, EDTA and blood glucose samples in the mixture, respectively. The analysis is done by taking this mixture as a measurand for glucose detection.

10.3 GEOMETRY FOR SENSING WITH INTEGRATED SoI WAVEGUIDES

Two of the basic waveguides which are used in glucose sensor are ridge waveguide and slot waveguide. Fig. 10.1(a) represents the structure of a fundamental ridge waveguide. This waveguide has two layers; a silica substrate layer at the bottom and a rectangular silicon ridge layer of height h and width w_r at the top of the waveguide. Even though ridge waveguide allows field confinement inside it but in the recent years, attempts have been made to confine the light in the lower refractive index region, and this can be attained by using slot waveguide as shown in Fig. 10.1(b). Two Si ridge waveguides of width w_{rs} and height h are placed leaving a gap of w_s, then a low refractive index material, SiO_2 is filled in this slot region. Slot waveguide provides strong electric field confinement in low index region.

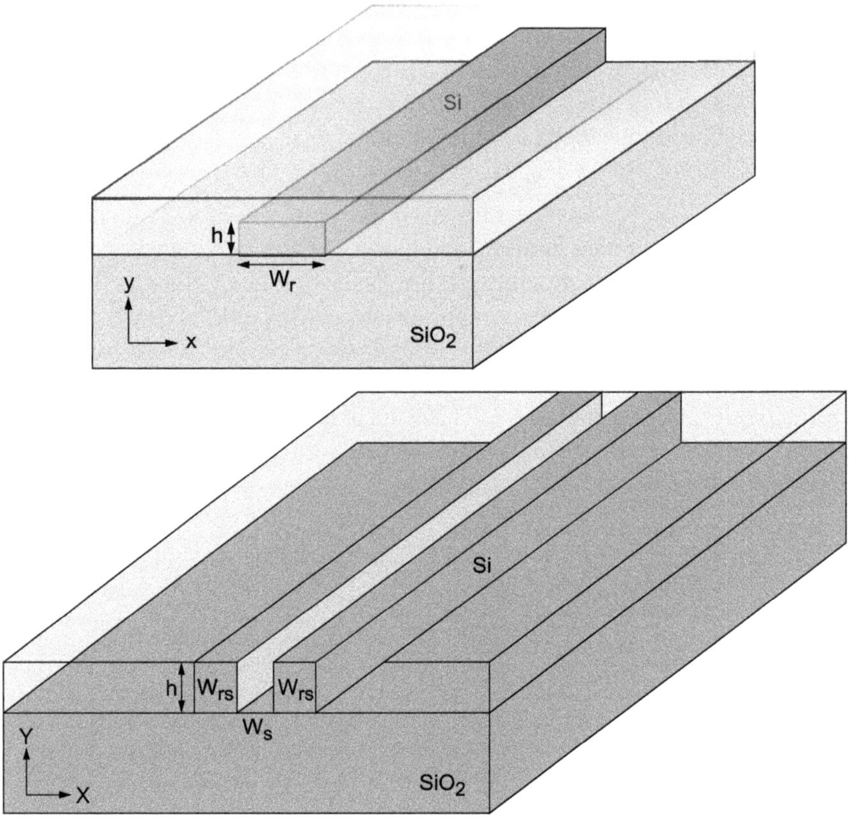

Fig. 10.1. Schematic of straight (a) ridge waveguide, and (b) slot waveguide.

Fig. 10.2(a) shows ridge waveguide based optical ring resonator, it consists of two parts; a Si bus waveguide of height h and width w_r along with Si ring waveguide of width w_r and ring radius r_r which is the distance from the center of ring to the center of circular waveguide. The spacing between bus waveguide and circular waveguide is represented by s_r. Fig. 10.2(b) shows the schematics of a slot waveguide based optical ring resonator. It consists of a straight bus waveguide as similar to previous case and the circular portion is made up of slot waveguide with r_s as ring radius from the center of the ring to center of the slot portion, while s_s is the spacing between bus waveguide and circular waveguide.

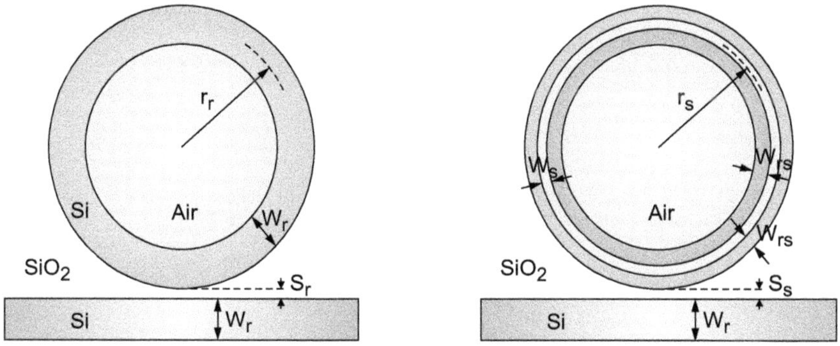

Fig. 10.2. Schematic of ring resonator glucose sensor with ridge bus waveguide and (a) ridge ring waveguide; (b) slot ring waveguide.

10.4 OPTIMIZATION OF SENSING GEOMETRY

Ridge waveguide with SiO_2 as substrate ($n_{SiO2} = 1.444$) and Si as guiding region ($n_{Si} = 3.476$) are taken to assure that this waveguide works on silicon-on-insulator phenomenon. Among standard height technologies of SOI waveguide fabrications, a waveguide height of 220 nm is taken in this work. In this case, waveguide width plays a crucial role of mode field confinement inside ridge region of SOI waveguide. Considering this, the ridge waveguide width needs to be optimized in the first place. To achieve this, air as the cladding material ($n_{air} = 1$) and operating wavelength of 1550 nm as mode analysis wavelength in frequency domain analysis of electromagnetic waves have been chosen. For this we have evaluated the values of confinement factor for a range of waveguide widths.

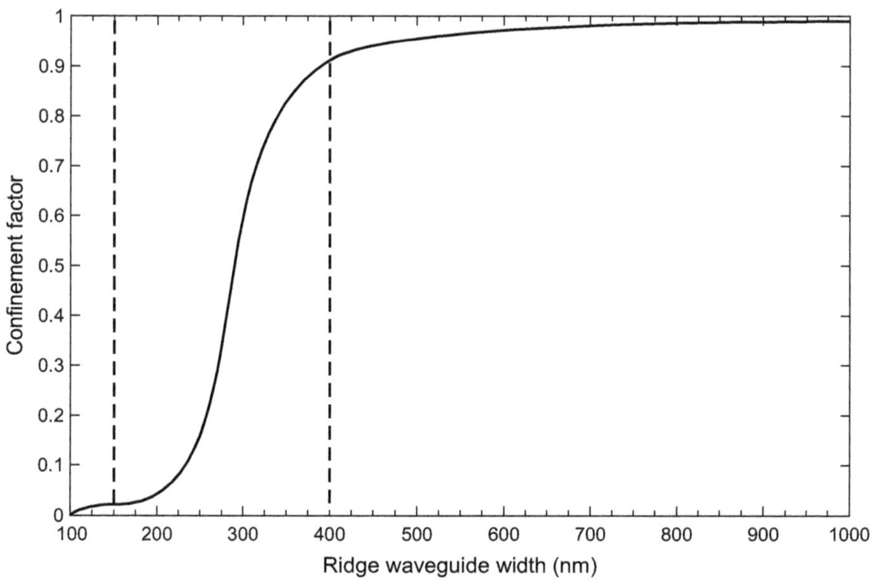

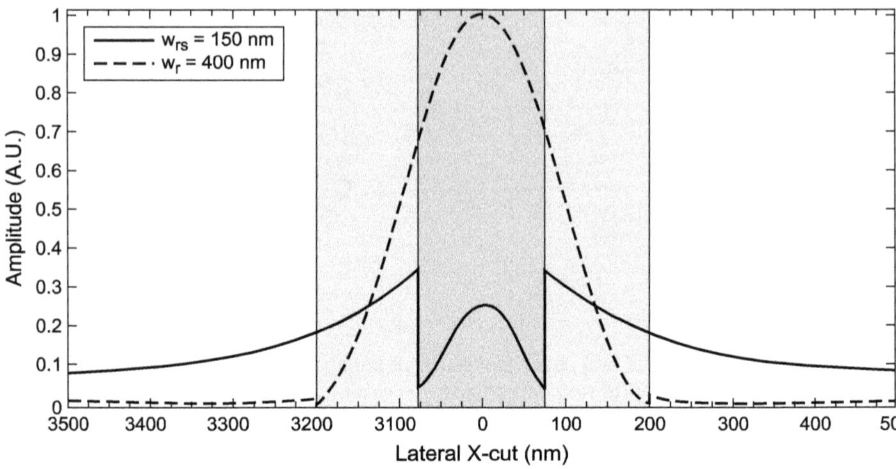

Fig. 10.3. Ridge bus waveguide (a) Confinement factor with variation in ridge waveguide width; (b) Mode field *x*-cut distribution profile for w_r = 150 nm and w_r = 400 nm.

Fig. 10.3(a) is plotted for power confinement factor in order to optimize the ridge waveguide width which is obtained very low for a ridge waveguide width less than 150 nm. However, it starts increasing as the width becomes more than 150 nm by following an increasing S-shape curve and nearly gets saturated after 400 nm. A constant free space wavenumber of 4.05×10^6 rad/m is obtained corresponding to all values of ridge waveguide width. A confinement factor of 0.02 (2%) and 0.9 (90%) are obtained at ridge waveguide width of 150 nm and 400 nm, respectively.

Fig. 10.3(b) explains the mode field distribution of ridge waveguide with ridge waveguide width (w_r) of 150 nm and 400 nm. In this figure, the profile obtained about mode distribution along the *x*-cut of ridge waveguide is the most compelling, which infers that, w_r = 400 nm supports the propagation of first order dominant quasi-TE mode inside silicon waveguide region. On the other hand, w_r = 150 nm

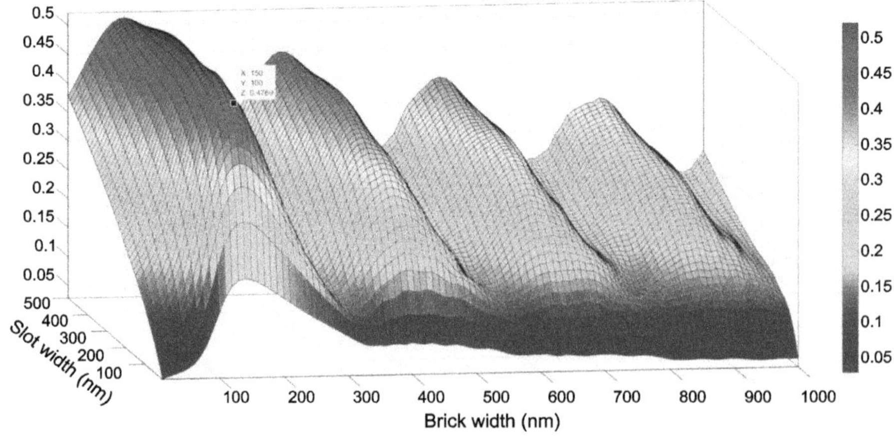

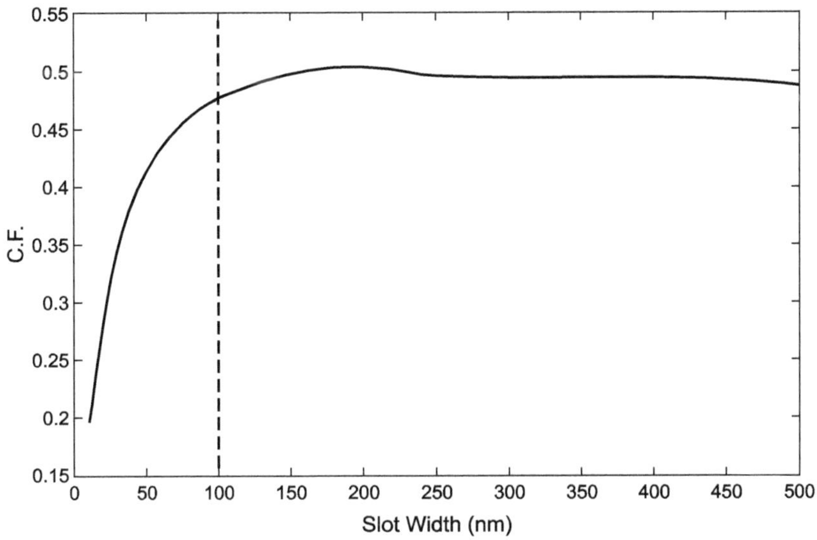

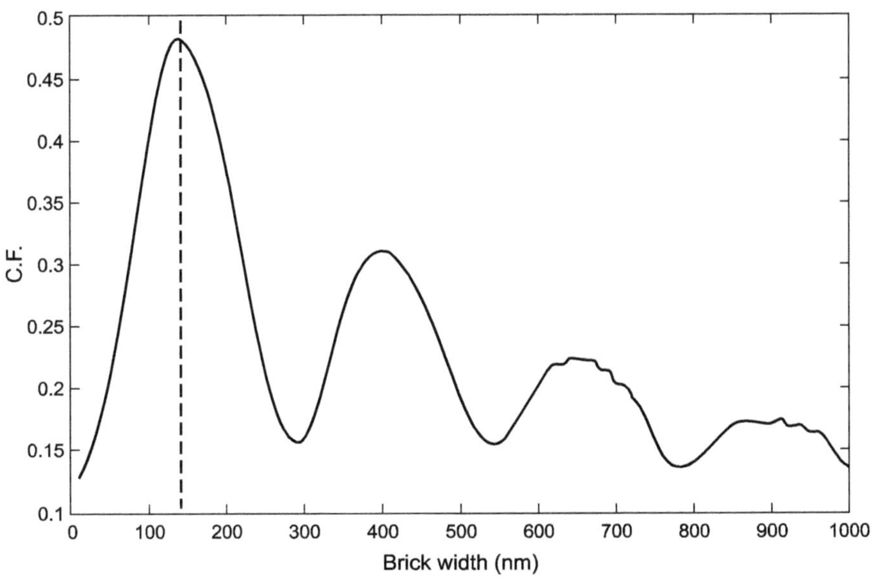

has field distribution lying majorly outside silicon waveguide region and thus evanescent field is obtained for this width which is evident from the mode profile curve. Observing these values and trend of the line graph in Fig. 10.3 infers that, a width of 400 nm is best suitable width for silicon brick in ridge waveguide and 150 nm silicon brick width is most suitable for the brick size in slot waveguide case. Since, w_r is the only parametric variable for ridge waveguide and its value of 400 nm is optimized for optimum performance. Therefore, another detailed analysis is required for slot waveguide optimization.

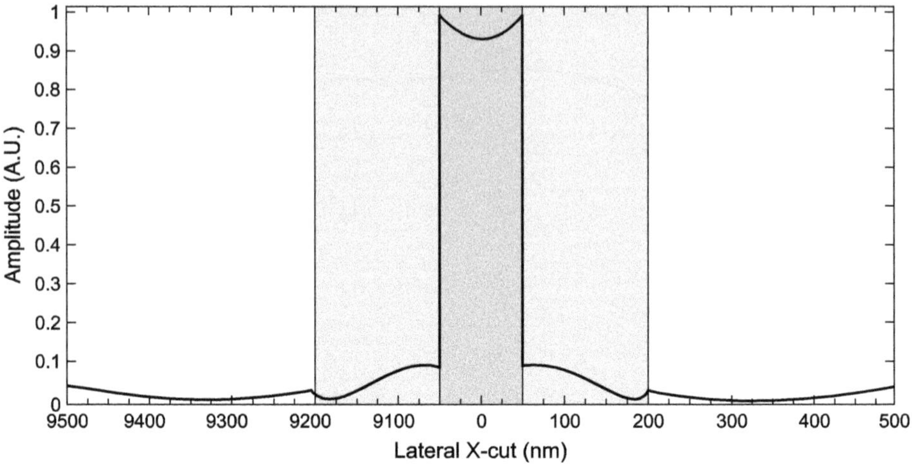

Fig. 10.4. (a) 2-variable parametric optimization of slot waveguide; (b) variation of C.F. with slot width (w_s) while keeping w_{rs} = 150 nm; (c) variation of C.F. with brick width (w_{rs}) while keeping w_s = 100 nm; (d) X-cut E-field distribution profile of slot waveguide for w_{rs} = 150 nm and w_s = 100 nm.

The width of silicon bricks in case of slot waveguide is optimized to 150 nm, even though the width of slot region needs to be optimized. So, the confinement factor inside the slot region of slot waveguide with SiO_2 as slot material is calculated simultaneously for a range of slot width (w_s) ranging from 10 nm to 500 nm and brick width ranging from 10 nm to 1000 nm as shown in Fig. 10.4(a). The maximum confinement is obtained in low refractive index slot region for taking slot width = 100 nm and rectangular brick width = 150 nm. Fig. 10.4(b) shows that after w_s = 100 nm, the confinement gets saturated to 47.69%, while the confinement increases from 20% to 47.69% for w_s between 10 nm to 100 nm. Thus, w_s = 100 nm is obtained as an optimum value of slot width in slot waveguide. Fig. 10.4(c) represents that the optimized width of rectangular waveguide region (w_{rs}) for slot waveguide is 150 nm and it is also justified from Fig. 10.4(a). The mode field distribution along X-cut plane of slot waveguide for w_{rs} = 150 nm and w_s = 100 nm is shown in Fig. 10.4(d).

10.5 RESULT AND DISCUSSION

While taking air as a cladding material, a wavelength of 1550 nm appeared to be one of the coupled resonance wavelengths travelling inside the ring. But when we replace this air cladding by blood sample as cladding material, the wavelengths get shifted a bit, resulting to the generation of new resonance wavelengths suitable to glucose sample. 1550 nm as a reference resonance wavelength is first taken and accordingly shift is calculated with reference to the replaced sample.

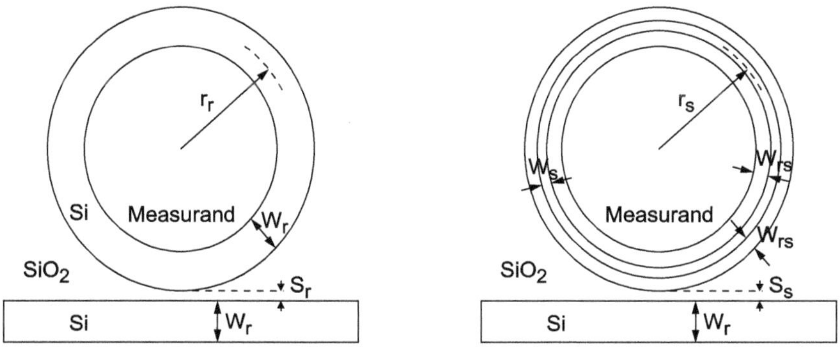

Fig. 10.5. Schematic of glucose sensing structure, (a) ridge ring waveguide with w_r = 400 nm; and (b) slot ring waveguide with w_{rs} = 150 nm and w_s = 100 nm.

10.5.1 Spectrum of Glucose Sensor

The graphs given in Fig. 10.6 to represent the new resonance wavelength shifted from 1550 nm to a new value, due to increase in glucose concentration from 10 mg/dl to 200 mg/dl. The graph is plotted for both the waveguide based ring resonators.

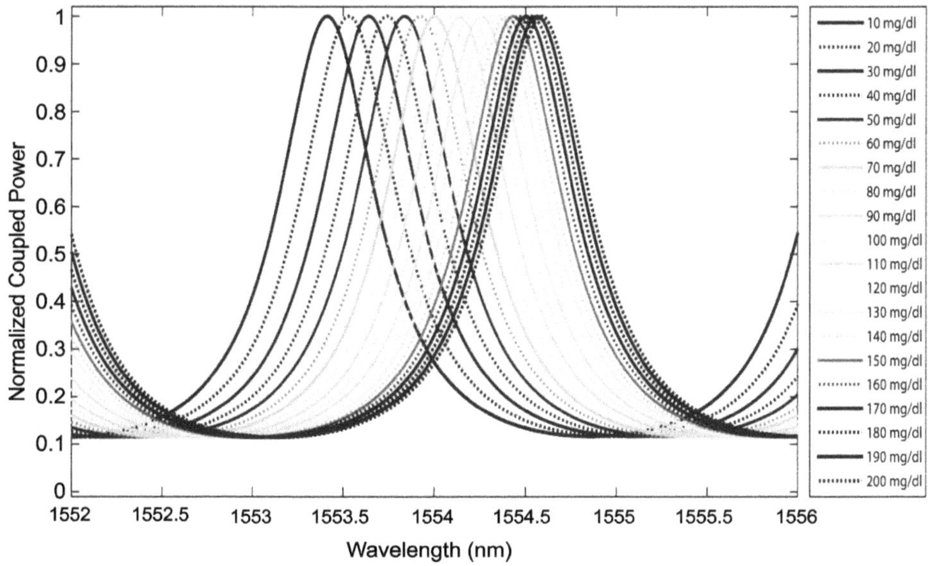

Fig. 10.5(a) and 10.1(b) shows the schematic of ring resonator with ridge ring and slot ring respectively. In Fig. 10.5, the cladding portion inside the ring is replaced with measurand in place of air with reference to Fig. 10.2. Considering this, the wavelength sweep is applied to input arm of straight ridge bus waveguide. Fig. 10.6(a) shows new resonance wavelength in case of ridge waveguide ring filled with measurand sample and while case of slot waveguide ring is shown in Fig. 10.6(b). A large wavelength shift to a new resonance value is observed in case

of slot waveguide than that of ridge waveguide based ring resonator for different glucose concentrations existing in blood sample. Higher the glucose concentration, more will be the wavelength shift away from 1550 nm, which is clear from Fig. 10.6.

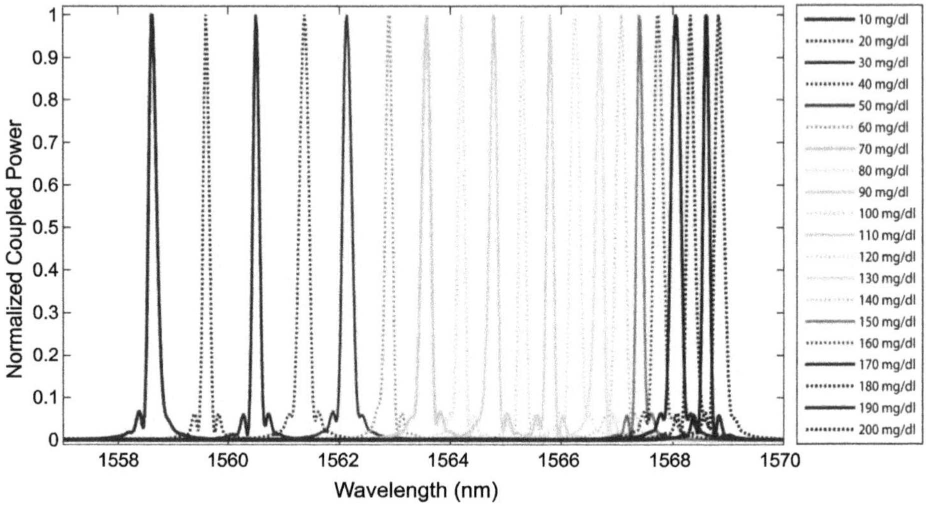

Fig. 10.6. Spectrum of glucose sensor, (a) ridge ring waveguide; and (b) slot ring waveguide.

10.5.2 Spectral Shift

Fig. 10.7 is plotted by combining the resulting resonance wavelengths for both the sensors. This is evident from the graph that the resonance wavelength shifts to the

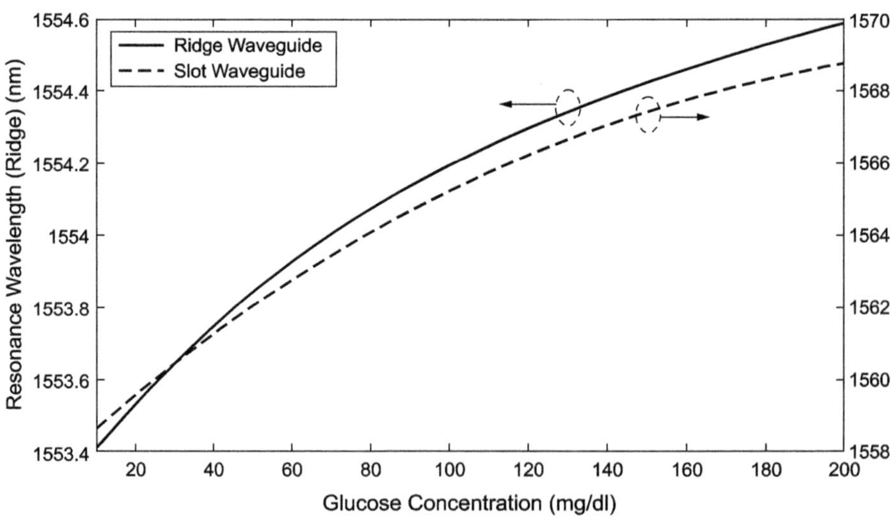

Fig. 10.7. Resonance wavelength for different glucose concentrations.

longer wavelength when the ring is made up of slot waveguide, as the glucose concentration existing in blood sample increases. For clinical aspects of this glucose sensor in detecting hyperglycemia and hypoglycemia, measurement points for 70 mg/dl, 110 mg/dl and 140 mg/dl should be known and well calibrated in the device. From Fig. 10.7, in case of ridge ring resonator, the resonance wavelength of 1554.00 nm, 1554.26 nm and 1554.40 nm are obtained for the glucose concentration of 70 mg/dl, 110 mg/dl and 140 mg/dl, respectively, while in case of slot ring resonator, resonance wavelength of 1563.58 nm, 1565.80 nm and 1567.07 nm are obtained.

The spectral shift indicates the shift in resonance wavelength from 1550 nm after changing the cladding material. This spectral shift $\Delta\lambda_r$ is calculated using

$$\Delta\lambda_r = \lambda_r - \lambda_0 \qquad \text{...(10.5)}$$

Where, $\lambda_0 = 1550$ nm (Resonance wavelength in the case of air cladding) and λ_r is the resonance wavelength due to glucose concentration.

This shift is calculated for a range (10 mg/dl – 200 mg/dl) of glucose concentration in blood sample. Fig. 10.8 shows that the spectral shift is greater while analysing slot waveguide based ring resonator sensor as compared to the sensor comprised of a ridge waveguide. Also, for low level of glucose concentration, the spectral shift in both the cases of ring is high and as the glucose concentration increases to approach 200 mg/dl concentration, the spectral shift in both the cases gets reduced. However, the spectral shift in case of slot ring waveguide is ~4 times more than that of ridge ring waveguide. The value of minimum shift in case of high glucose concentration between 190 mg/dl and 200 mg/dl for slot waveguide based structure is approximately 0.5 nm, which can be detected by latest high resolution spectrum analyser, but the ridge waveguide based sensor provides minimum wavelength shift nearly equal to 0.05 nm which is very difficult to analyse practically.

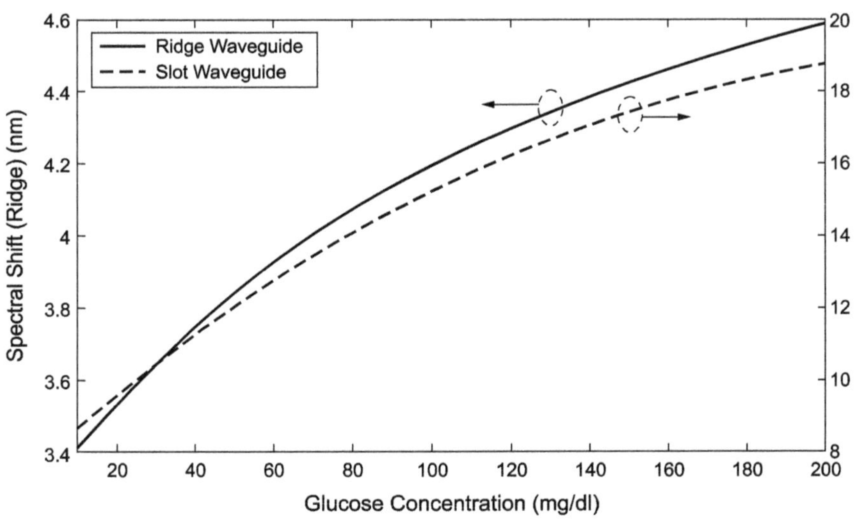

Fig. 10.8. Spectral shift ($\Delta\lambda_r$) for different glucose concentrations.

10.5.3 Sensitivity for Different Glucose Concentrations

Wavelength sensitivity is one of the desired characteristics to be derived for a sensor. It shows how much a sensor is sensitive to the wavelength. This can be evaluated as:

$$S_\lambda = \frac{\Delta\lambda_r}{\Delta c} \quad ...(10.6)$$

Fig. 10.9 reflects the waveguide sensitivity as a function of glucose concentration. Slot based ring waveguide is more sensitive to glucose concentration as compared to ridge based waveguide. Thus, it is inferred that small amount of glucose change in blood can be detected using slot ring resonator. In order to achieve a better sensitivity, ridge based sensor is replaced with slot waveguide based sensor as it is more sensitive to the change in glucose concentration.

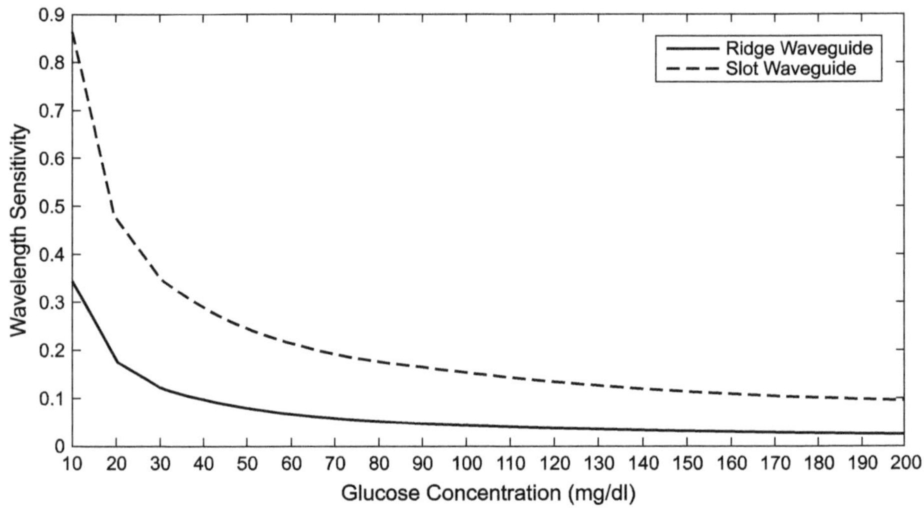

Fig. 10.9. Waveguide sensitivity (S_λ) for different glucose concentrations.

Index sensitivity of the device is also a desired characteristic to be acquired for a sensor. It shows how much a sensor is sensitive to change in refractive index of measurand (Barrios *et al.*, 2007, 2008; Claes *et al.*, 2009). Figure 10.10 shows that the resonance wavelength redshifts with increasing refractive index units (RIU) with an average slope of 60 nm/RIU for ridge ring and 360 nm/RIU for slot ring respectively. A sensitivity of 60 nm/RIU is obtained for ridge waveguide based ring resonator, whereas we can obtain 70 nm/RIU (De Vos *et al.*, 2007) practically from biosensors based on ridge ring waveguide. On the other hand, the sensitivity of 360 nm/RIU is obtained for slot waveguide based ring resonator, which is comparable with practically obtained values 298 nm/RIU (Claes *et al.*, 2009) and 212 nm/RIU (Barrios *et al.*, 2007). The value of sensitivity for slot waveguide is six times higher than exhibited by conventional ridge waveguide. These values are favorably comparable with those of state-of-the-art integrated photonic biosensors based on conventional slot and ridge waveguides, demonstrating the benefits of using slot waveguides to detect small glucose concentration change.

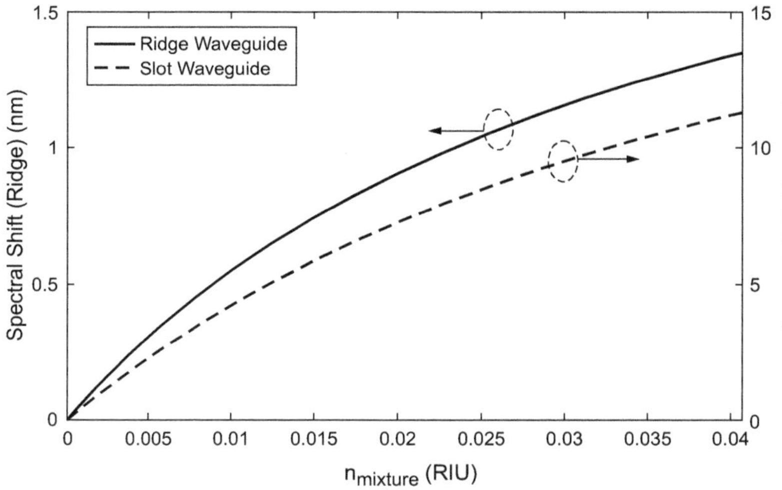

Fig. 10.10. Spectral shift plotted as a function of ambient index ($\Delta n_{mixture}$).

10.5.4 Fabrication Tolerance Analysis

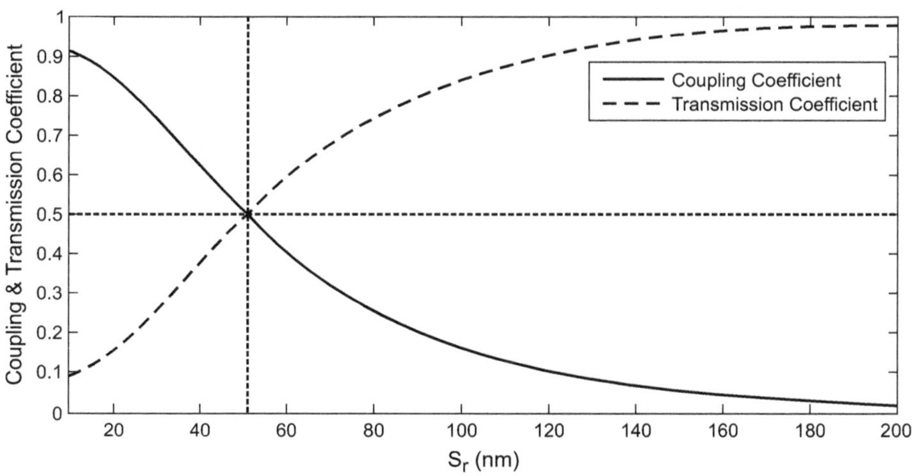

Tolerance limit is been evaluated in relation with the spacing between the edge of circular waveguide and straight waveguide for fabrication point of view. This spacing will affect the coupling efficiency of the waveguides. While fabricating the structure using ridge waveguide in the circular ring portion, up to 50 nm (S_r) spacing in between the waveguides is tolerable because up to 50 nm spacing, the power being coupled inside the ring is at least 3dB and it is represented by Fig. 10.11(a).

However, beyond this spacing range, coupling power will deteriorate below 3dB. Likewise, the spacing preferred for slot waveguide based ring is nearly 2 nm and it is evident from Fig. 10.11(b). The spacing value within this specified range is marked. All of these values will provide tolerable spacing between the waveguides for fabrication point of view.

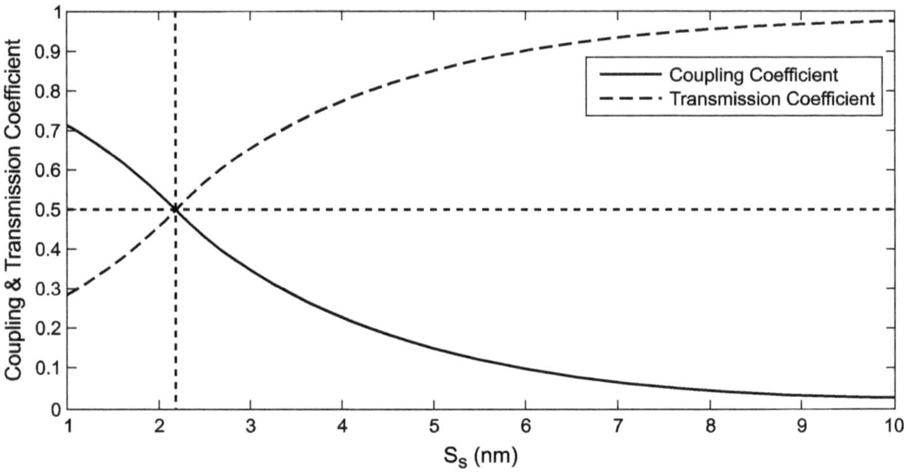

Fig. 10.11. Coupling and transmission coefficient with spacing between bus and ring waveguide to indicate fabrication tolerance for (a) ridge ring waveguide; and (b) slot ring waveguide.

10.5.5 Quality Factor for Different Glucose Concentrations

To realize the complete designing of this ring resonator based sensor, the significant parameters of the ring resonator are evaluated using analytical formulae (Bogaerts

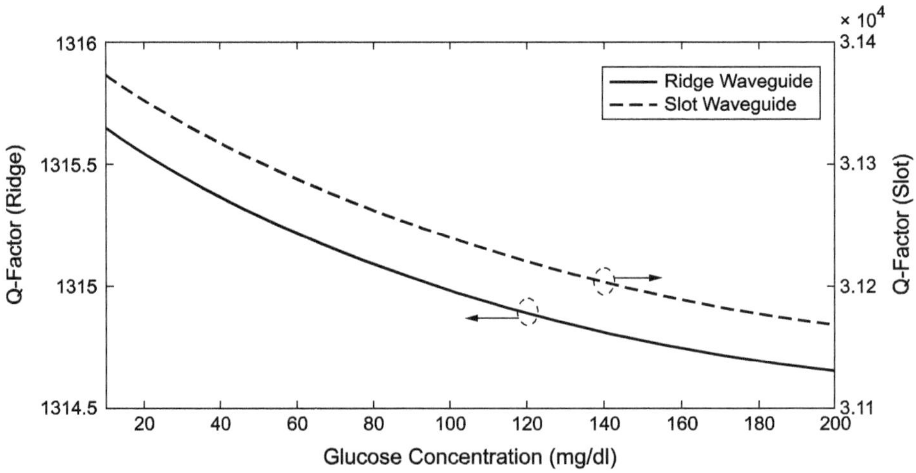

Fig. 10.12. Quality factor for different glucose concentrations.

et al., 2012). The parameter quality-factor (Q) varies with changing concentration as shown in Fig. 10.12. The Q-value is comparatively very high for the slot waveguide based sensor. However, both the sensors follow the same trend i.e. it decreases with the increasing value of glucose concentration.

In case of ridge ring resonator, the values of FSR, FWHM and Finesse for air filled cladding as well as for cladding filled with different glucose concentrations are 3 nm, 0.6807 nm and 4.407, respectively. However, in case of slot ring resonator, the value of FSR and finesse increase to 20 nm and 111.110, but the value of FWHM decreases to 0.18 nm. Higher value of FSR, smaller FWHM and higher Finesse values are being obtained in the case of ring incorporated with slot waveguide is preferable for sensing application as high value of FSR corresponds with longer wavelength shift. The behavior of slot waveguide based ring is better than ridge ring for the purpose of sensing which is justified from these discussions.

10.6 SUMMARY FOR RING RESONATOR SENSING APPLICATION

Over the last few decades, significant efforts have been done towards the development of photonic glucose sensor, so that continuous monitoring of hyperglycaemia and hypoglycaemia can be done efficiently. This chapter presents a minimal invasive approach performing homogeneous glucose sensing by an optical ring resonator to detect the glucose level in blood sample. Two different designs of the sensor is been evaluated. In the first sensor, the ring waveguide is constructed using a ridge waveguide and in the other one, slot waveguide in ring section is used for enhancing the performance of the sensor. The optimized value of ring radius is calculated as 35.3μm for ridge ring and 33.5μm for slot ring resonator. The coupling waveguide is ridge waveguide in both the cases of ridge and slot ring, having the optimal dimension of $w_r = 400$ nm and $h = 220$ nm. The important characteristics of the sensor such as sensitivity, resonant wavelength, spectral shift, and Q-factor are plotted for both the optimized ring waveguide designs to have a comparative study of the characteristics. The key waveguide design parameters of ridge and slot waveguide are optimized for both straight bus waveguide and ring resonator cases to achieve its highest efficient propagation with confinement of 47.69%. After observing the results, it is concluded that slot waveguide based ring resonator shows longer wavelength shift and higher average sensitivity, so it proves to be a better sensor that provides fast and accurate result.

❑❑❑

References

REFERENCES

- Aalto, T.T. *et al.* (2004) 'Development of silicon-on-insulator waveguide technology', in Y. Sidorin and A. Tervonen (eds), p. 81. Available at: https://doi.org/10.1117/12.537540.
- Abdul Rashid, J.I. *et al.* (2013) 'The Development of Silicon Nanowire as Sensing Material and Its Applications', *Journal of Nanomaterials*, 2013, pp. 1–16. Available at: https://doi.org/10.1155/2013/328093.
- Agrawal, G.P. (2007) *Nonlinear Fiber Optics*. 4 ed. Academic.
- Agrawal, G.P. (2010) *Fiber-Optic Communication Systems*. 4th edn. Oxford: Wiley-Blackwell.
- Alarousu, E. *et al.* (2004) '<title>Noninvasive glucose sensing in scattering media using OCT, PAS, and TOF techniques</title>', in V. V. Tuchin (ed.), pp. 33–41. Available at: https://doi.org/10.1117/12.578321.
- Almeida, V.R. *et al.* (2004) 'Guiding and confining light in void nanostructure', *Optics Letters*, 29(11), p. 1209. Available at: https://doi.org/10.1364/OL.29.001209.
- An, G. *et al.* (2017) 'Glucose sensor realized with photonic crystal fiber-based Sagnac interferometer', *Optics Communications*, 405, pp. 143–146. Available at: https://doi.org/10.1016/j.optcom.2017.08.003.
- An, L. *et al.* (2014) 'Wavelength conversion in highly nonlinear silicon–organic hybrid slot waveguides', *Applied Optics*, 53(22), p. 4886. Available at: https://doi.org/10.1364/AO.53.004886.
- Anderson, P.A., Schmidt, B.S. and Lipson, M. (2006) 'High confinement in silicon slot waveguides with sharp bends', *Optics Express*, 14(20), p. 9197. Available at: https://doi.org/10.1364/OE.14.009197.
- Arti Agrawal, B.M.A.R. (2013) *Finite Element Modeling Methods for Photonics*. Artech House.
- Ashok, N., Lee, Y.L. and Shin, W. (2016) 'Design and Study of Strip-Slot Waveguide Structure for Dispersion Analysis', *IEEE Photonics Journal*, 8(1), pp. 1–8. Available at: https://doi.org/10.1109/JPHOT.2016.2519285.

- Azuelos, P. *et al.* (2017) 'High sensitivity optical biosensor based on polymer materials and using the Vernier effect', *Optics Express*, 25(24), p. 30799. Available at: https://doi.org/10.1364/OE.25.030799.
- Bae, J. *et al.* (2010) 'Si nanowire metal–insulator–semiconductor photodetectors as efficient light harvesters', *Nanotechnology*, 21(9), p. 095502. Available at: https://doi.org/10.1088/0957-4484/21/9/095502.
- Baehr-Jones, T. *et al.* (2004) 'High-Q ring resonators in thin silicon-on-insulator', *Applied Phy*Sics *Letters*, 85(16), pp. 3346–3347. Available at: https://doi.org/10.1063/1.1781355.
- Bardeen, J. (1988) 'Solid State Physics—1947', *Microelectronics International*, 5(3), pp. 6–7. Available at: https://doi.org/10.1108/eb044332.
- Barozzi, M. *et al.* (2017) 'Optical Fiber Sensors for Label-Free DNA Detection', *Journal of Lightwave Technology*, 35(16), pp. 3461–3472. Available at: https://doi.org/10.1109/JLT.2016.2607024.
- Barrios, C.A. *et al.* (2007) 'Slot-waveguide biochemical sensor', *Optics Letters*, 32(21), p. 3080. Available at: https://doi.org/10.1364/OL.32.003080.
- Barrios, C.A. *et al.* (2008) 'Label-free optical biosensing with slot-waveguides', *Optics Letters*, 33(7), p. 708. Available at: https://doi.org/10.1364/OL.33.000708.
- Barrios, C.A. and Lipson, M. (2005) 'Electrically driven silicon resonant light emitting device based on slot-waveguide', *Optics Express*, 13(25), p. 10092. Available at: https://doi.org/10.1364/OPEX.13.010092.
- Bhardwaj, V. and Singh, V.K. (2016) 'Fabrication and characterization of cascaded tapered Mach-Zehnder interferometer for refractive index sensing', *Sensors and Actuators A: Phy*Sical, 244, pp. 30–34. Available at: https://doi.org/10.1016/j.sna.2016.04.008.
- Bhushan, A.S., Coppinger, F. and Jalali, B. (1998) 'Time-stretched analogue-to-digital conversion', *Electronics Letters*, 34(9), p. 839. Available at: https://doi.org/10.1049/el:19980629.
- Boeuf, F. *et al.* (2015) 'Recent Progress in Silicon Photonics R&D and Manufacturing on 300mm Wafer Platform', in *Optical Fiber Communication Conference*. Washington, D.C.: OSA, p. W3A.1. Available at: https://doi.org/10.1364/OFC.2015.W3A.1.
- Bogaerts, W. *et al.* (2012) 'Silicon microring resonators', *Laser & Photonics Reviews*, 6(1), pp. 47–73. Available at: https://doi.org/10.1002/lpor.201100017.
- Cam Hoang, T.H. (2022) 'Analysis on slotted photonic crystal cavity and waveguide combination in silicon-on-insulator platform', *Optik*, 251, p. 168465. Available at: https://doi.org/10.1016/j.ijleo.2021.168465.

- Chan, C.K. *et al.* (2008) 'High-performance lithium battery anodes using silicon nanowires', *Nature Nanotechnology*, 3(1), pp. 31–35. Available at: https://doi.org/10.1038/nnano.2007.411.
- Chen, X. *et al.* (2013) 'Nanowire-based gas sensors', *Sensors and Actuators B: Chemical*, 177, pp. 178–195. Available at: https://doi.org/10.1016/j.snb.2012.10.134.
- Chung, Y. and Dagli, N. (1990) 'An assessment of finite difference beam propagation method', *IEEE Journal of Quantum Electronics*, 26(8), pp. 1335–1339. Available at: https://doi.org/10.1109/3.59679.
- Ciminelli, C. *et al.* (2013) 'Label-free optical resonant sensors for biochemical applications', *Progress in Quantum Electronics*, 37(2), pp. 51–107. Available at: https://doi.org/10.1016/j.pquantelec.2013.02.001.
- Claes, T. *et al.* (2009) 'Label-Free Biosensing With a Slot-Waveguide-Based Ring Resonator in Silicon on Insulator', *IEEE Photonics Journal*, 1(3), pp. 197–204. Available at: https://doi.org/10.1109/JPHOT.2009.2031596.
- Coldren, L.A. *et al.* (2004) 'Tunable Semiconductor Lasers: A Tutorial', *Journal of Lightwave Technology*, 22(1), pp. 193–202. Available at: https://doi.org/10.1109/JLT.2003.822207.
- Cooper, M.A. (ed.) (2009) *Label-Free Biosensors*. Cambridge: Cambridge University Press. Available at: https://doi.org/10.1017/CBO9780511626531.
- Cui, Y. (2001) 'Nanowire Nanosensors for Highly Sensitive and Selective Detection of Biological and Chemical Species', *Science*, 293(5533), pp. 1289–1292. Available at: https://doi.org/10.1126/science.1062711.
- Dar, T. *et al.* (2012) 'Label-free slot-waveguide biosensor for the detection of DNA hybridization', *Applied Optics*, 51(34), p. 8195. Available at: https://doi.org/10.1364/AO.51.008195.
- Dattner, Y. and Yadid-Pecht, O. (2011) 'Analysis of the Effective Refractive Index of Silicon Waveguides Through the Constructive and Destructive Interference in a Mach–Zehnder Interferometer', *IEEE Photonics Journal*, 3(6), pp. 1123–1132. Available at: https://doi.org/10.1109/JPHOT.2011.2171678.
- Dell'Olio, Conteduca, De Palo, C. (2017) 'Design of a New Ultracompact Resonant Plasmonic Multi-Analyte Label-Free Biosensing Platform', *Sensors*, 17(12), p. 1810. Available at: https://doi.org/10.3390/s17081810.
- Dell'Olio, F. *et al.* (2015) 'New ultrasensitive resonant photonic platform for label-free biosensing', *Optics Express*, 23(22), p. 28593. Available at: https://doi.org/10.1364/OE.23.028593.
- Dell'Olio, F. and Passaro, V.M. (2007) 'Optical sensing by optimized silicon slot waveguides', *Optics Express*, 15(8), p. 4977. Available at: https://doi.org/10.1364/OE.15.004977.

- 'Diagnosis and Classification of Diabetes Mellitus' (2014) *Diabetes Care*, 37(Supplement_1), pp. S81–S90. Available at: https://doi.org/10.2337/dc14-S081.
- Dick, K.A. *et al.* (2005) 'Failure of the Vapor–Liquid–Solid Mechanism in Au-Assisted MOVPE Growth of InAs Nanowires', *Nano Letters*, 5(4), pp. 761–764. Available at: https://doi.org/10.1021/nl050301c.
- Elhadj, S., Singh, G. and Saraf, R.F. (2004) 'Optical properties of an immobilized DNA monolayer from 255 to 700 nm.', *Langmuir : the ACS journal of surfaces and colloids*, 20(13), pp. 5539–43. Available at: https://doi.org/10.1021/la049653+.
- Eti, N. and Kurt, H. (2016) 'Model Analysis of Ridge and Rib Types of Silicon Waveguides With Void Compositions', *IEEE Journal of Quantum Electronics*, 52(10), pp. 1–7. Available at: https://doi.org/10.1109/JQE.2016.2598082.
- Fathi Aghdam, F., Liao, H. and Huang, Q. (2017) 'Modeling Interaction in Nanowire Growth Process Toward Improved Yield', *IEEE Transactions on Automation Science and Engineering*, 14(2), pp. 1139–1149. Available at: https://doi.org/10.1109/TASE.2015.2499210.
- Fischer, U., ZInke, T. and Petermann, K. (1995) 'Integrated optical waveguide switches in SOI', in 1995 *IEEE International SOI Conference Proceedings*. IEEE, pp. 141–142. Available at: https://doi.org/10.1109/SOI.1995.526500.
- Flueckiger, J. *et al.* (2016) 'Sub-wavelength grating for enhanced ring resonator biosensor', *Optics Express*, 24(14), p. 15672. Available at: https://doi.org/10.1364/OE.24.015672.
- Foster, M.A. *et al.* (2006) 'Broad-band optical parametric gain on a silicon photonic chip', *Nature*, 441(7096), pp. 960–963. Available at: https://doi.org/10.1038/nature04932.
- Fu, Y.Q. *et al.* (2009) 'Deep reactive ion etching as a tool for nanostructure fabrication', *Journal of Vacuum Science & Technology B: Microelectronics and Nanometer Structures*, 27(3), p. 1520. Available at: https://doi.org/10.1116/1.3065991.
- Fuhrmann, B. *et al.* (2005) 'Ordered Arrays of Silicon Nanowires Produced by Nanosphere Lithography and Molecular Beam Epitaxy', *Nano Letters*, 5(12), pp. 2524–2527. Available at: https://doi.org/10.1021/nl051856a.
- Garnett, E. and Yang, P. (2010) 'Light Trapping in Silicon Nanowire Solar Cells', *Nano Letters*, 10(3), pp. 1082–1087. Available at: https://doi.org/10.1021/nl100161z.
- Gehlot, K. and Sharma, A. (2014) 'Approximate analysis of planar photonic bandgap waveguides: A simple semi-analytical method', *Optical and Quantum Electronics*, 46(3), pp. 455–464. Available at: https://doi.org/10.1007/s11082-014-9873-7.

- Gelfand, R.M. *et al.* (2011) 'Towards an Integrated Chip-Scale Plasmonic Biosensor', *Optics and Photonics News*, 22(4), p. 32. Available at: https://doi.org/10.1364/OPN.22.4.000032.
- Geraghty, D.F. *et al.* (2008) 'A Simplified Optical Correlator and Its Application to Packet-Header Recognition', *IEEE Photonics Technology Letters*, 20(7), pp. 487–489. Available at: https://doi.org/10.1109/LPT.2008.918218.
- Ghatak, A.K. and Thyagarajan, K. (1989) *Optical Electronics*. Cambridge: Cambridge University Press. Available at: https://doi.org/10.1017/CBO9781139167857.
- Goldberger, J. *et al.* (2006) 'Silicon Vertically Integrated Nanowire Field Effect Transistors', *Nano Letters*, 6(5), pp. 973–977. Available at: https://doi.org/10.1021/nl060166j.
- Gough, D.A. and Armour, J.C. (1995) 'Development of the Implantable Glucose Sensor: What Are the Prospects and Why Is It Taking So Long?', *Diabetes*, 44(9), pp. 1005–1009. Available at: https://doi.org/10.2337/diab.44.9.1005.
- Guan, X. *et al.* (2014) 'Low-loss ultracompact transverse-magnetic-pass polarizer with a silicon subwavelength grating waveguide', *Optics Letters*, 39(15), p. 4514. Available at: https://doi.org/10.1364/OL.39.004514.
- Gunasundari, E. *et al.* (2014) 'Waveguiding properties of a silicon nanowire embedded photonic crystal fiber', *Optical Materials*, 36(5), pp. 958–964. Available at: https://doi.org/10.1016/j.optmat.2013.12.048.
- Guo, X. and Tong, L. (2008) 'Supported microfiber loops for optical sensing', *Optics Express*, 16(19), p. 14429. Available at: https://doi.org/10.1364/OE.16.014429.
- Guo, Y. *et al.* (2016) 'Bilayer dispersion-flattened waveguides with four zero-dispersion wavelengths', *Optics Letters*, 41(21), p. 4939. Available at: https://doi.org/10.1364/OL.41.004939.
- Halir, R. *et al.* (2012) 'Colorless directional coupler with dispersion engineered sub-wavelength structure', *Optics Express*, 20(12), p. 13470. Available at: https://doi.org/10.1364/OE.20.013470.
- Halir, R. *et al.* (2015) 'Waveguide sub-wavelength structures: a review of principles and applications', *Laser & Photonics Reviews*, 9(1), pp. 25–49. Available at: https://doi.org/10.1002/lpor.201400083.
- Han, H., Huang, Z. and Lee, W. (2014) 'Metal-assisted chemical etching of silicon and nanotechnology applications', *Nano Today*, 9(3), pp. 271–304. Available at: https://doi.org/10.1016/j.nantod.2014.04.013.
- Han, Y. *et al.* (2020) 'Bufferless 15 μm III-V lasers grown on Si-photonics 220 nm silicon-on-insulator platforms', *Optica*, 7(2), p. 148. Available at: https://doi.org/10.1364/OPTICA.381745.

- Hochberg, M. and Baehr-Jones, T. (2010) 'Towards fabless silicon photonics', *Nature Photonics*, 4(8), pp. 492–494. Available at: https://doi.org/10.1038/nphoton.2010.172.
- Holonyak, N. (1997) 'The semiconductor laser: a thirty-five-year perspective', *Proceedings of the IEEE*, 85(11), pp. 1678–1693. Available at: https://doi.org/10.1109/5.649645.
- Holonyak, N. and Bevacqua, S.F. (1962) 'COHERENT (VISIBLE) LIGHT EMISSION FROM Ga(As 1 − x P x) JUNCTIONS', *Applied PhySics Letters*, 1(4), pp. 82–83. Available at: https://doi.org/10.1063/1.1753706.
- Hong, J., Huang, W. and Makino, T. (1992) 'On the transfer matrix method for distributed-feedback waveguide devices', *Journal of Lightwave Technology*, 10(12), pp. 1860–1868. Available at: https://doi.org/10.1109/50.202840.
- Hsin, C.-L., Wu, M.-H. and Wang, W.-C. (2019) 'Thermoelectric Devices by Half-Millimeter-Long Silicon Nanowires Arrays', *IEEE Transactions on Nanotechnology*, 18, pp. 921–924. Available at: https://doi.org/10.1109/TNANO.2019.2938624.
- Humayun, M.T. *et al.* (2016) 'Ubiquitous Low-Cost Functionalized Multi-Walled Carbon Nanotube Sensors for Distributed Methane Leak Detection', *IEEE Sensors Journal*, 16(24), pp. 8692–8699. Available at: https://doi.org/10.1109/JSEN.2016.2581832.
- *Integrated Ring Resonators* (2007). Berlin, Heidelberg: Springer Berlin Heidelberg (Springer Series in Optical Sciences). Available at: https://doi.org/10.1007/978-3-540-68788-7.
- Iqbal, M. (2014) 'Dispersion modified slot optical waveguides', *Optik - International Journal for Light and Electron Optics*, 125(14), pp. 3549–3554. Available at: https://doi.org/10.1016/j.ijleo.2014.01.053.
- Iqbal, M. *et al.* (2020) 'Designing Optical Waveguides: Myth and Reality', *Brazilian Journal of PhySics*, 50(6), pp. 857–873. Available at: https://doi.org/10.1007/s13538-020-00763-w.
- Jafari, Z. *et al.* (2016) 'Parameter Space Exploration in Dispersion Engineering of Multilayer Silicon Waveguides from Near-Infrared to Mid-Infrared', *Journal of Lightwave Technology*, 34(16), pp. 3696–3702. Available at: https://doi.org/10.1109/JLT.2016.2581845.
- Jalali, B. *et al.* (2006) 'Prospects for Silicon Mid-IR Raman Lasers', *IEEE Journal of Selected Topics in Quantum Electronics*, 12(6), pp. 1618–1627. Available at: https://doi.org/10.1109/JSTQE.2006.885340.
- Jalali, B., Paniccia, M. and Reed, G. (2006) 'Silicon photonics', *IEEE Microwave Magazine*, 7(3), pp. 58–68. Available at: https://doi.org/10.1109/MMW.2006.1638290.

- Jiasheng Ni *et al.* (2008) 'Fiber methane gas sensor and its application in methane outburst prediction in coal mine', in 2008 *1st ASia-Pacific Optical Fiber Sensors Conference*. IEEE, pp. 1–4. Available at: https://doi.org/10.1109/APOS.2008.5226294.
- Jones, R. *et al.* (2005) 'Net continuous wave optical gain in a low loss silicon-on-insulator waveguide by stimulated Raman scattering', *Optics Express*, 13(2), p. 519. Available at: https://doi.org/10.1364/OPEX.13.000519.
- Kaminow, I.P. (2008) 'Optical Integrated Circuits: A Personal Perspective', *Journal of Lightwave Technology*, 26(9), pp. 994–1004. Available at: https://doi.org/10.1109/JLT.2008.922149.
- Kao, K.C. and Hockham, G.A. (1966) 'Dielectric-fibre surface waveguides for optical frequencies', *Proceedings of the Institution of Electrical Engineers*, 113(7), pp. 1151–1158. Available at: https://doi.org/10.1049/piee.1966.0189.
- Kapoor, A. and Singh, G.S. (2000) 'Mode classification in cylindrical dielectric waveguides', *Journal of Lightwave Technology*, 18(6), pp. 849–852. Available at: https://doi.org/10.1109/50.848397.
- Kargar, A. and Chao, C.-Y. (2011) 'Design and optimization of waveguide sensitivity in slot microring sensors', *Journal of the Optical Society of America A*, 28(4), p. 596. Available at: https://doi.org/10.1364/JOSAA.28.000596.
- Kazanskiy, N.L., Butt, M.A. and Khonina, S.N. (2021) 'Silicon photonic devices realized on refractive index engineered subwavelength grating waveguides-A review', *Optics & Laser Technology*, 138, p. 106863. Available at: https://doi.org/10.1016/j.optlastec.2020.106863.
- Khalil, O.S. (1999) 'Spectroscopic and Clinical Aspects of Noninvasive Glucose Measurements', *Clinical Chemistry*, 45(2), pp. 165 LP – 177. Available at: http://clinchem.aaccjnls.org/content/45/2/165.abstract.
- Khandokar, R.H. *et al.* (2016) 'Performance of Planar, Rib, and Photonic Crystal Silicon Waveguides in Tailoring Group-Velocity Dispersion and Mode Loss', *IEEE Journal of Selected Topics in Quantum Electronics*, 22(2), pp. 73–80. Available at: https://doi.org/10.1109/JSTQE.2015.2479359.
- Khanna, A. *et al.* (2009) 'Control of optical mode properties in cross-slot waveguides', *Applied Optics*, 48(34), p. 6547. Available at: https://doi.org/10.1364/AO.48.006547.
- Khodadad, I. *et al.* (2014) 'Optimization of multiple-slot waveguides for biochemical sensing', *Applied Optics*, 53(23), p. 5169. Available at: https://doi.org/10.1364/AO.53.005169.
- Khorasaninejad, M. *et al.* (2011) 'Optical bio-chemical sensors on SNOW ring resonators', *Optics Express*, 19(18), p. 17575. Available at: https://doi.org/10.1364/OE.19.017575.

- Khorasaninejad, M. and Saini, S.S. (2010) 'Silicon nanowire optical waveguide (SNOW)', *Optics Express*, 18(22), p. 23442. Available at: https://doi.org/10.1364/OE.18.023442.
- Khorasaninejad, M. and Saini, S.S. (2011) 'Bend Waveguides on Silicon Nanowire Optical Waveguide (SNOW)', *IEEE Photonics Journal*, 3(4), pp. 696–702. Available at: https://doi.org/10.1109/JPHOT.2011.2160527.
- Kish, F. *et al.* (2018) 'System-on-Chip Photonic Integrated Circuits', *IEEE Journal of Selected Topics in Quantum Electronics*, 24(1), pp. 1–20. Available at: https://doi.org/10.1109/JSTQE.2017.2717863.
- Kogelnik, H. (1975) 'Theory of Dielectric Waveguides', in, pp. 13–81. Available at: https://doi.org/10.1007/978-3-662-43208-2_2.
- Koschinsky, T. and Heinemann, L. (2001) 'Sensors for glucose monitoring: technical and clinical aspects', *Diabetes/Metabolism Research and Reviews*, 17(2), pp. 113–123. Available at: https://doi.org/10.1002/dmrr.188.
- Kozma, P. *et al.* (2014) 'Integrated planar optical waveguide interferometer biosensors: A comparative review', *Biosensors and Bioelectronics*, 58, pp. 287–307. Available at: https://doi.org/10.1016/j.bios.2014.02.049.
- Kumar, V. (2010) 'Fiber optic methane and strain sensors for mines', in *International Conference On Photonics* 2010. IEEE, pp. 1–4. Available at: https://doi.org/10.1109/ICP.2010.5604439.
- Kumari, B., Varshney, R.K. and Pal, B.P. (2018) 'Design of chip scale silicon rib slot waveguide for sub-ppm detection of N 2 O gas at mid-IR band', *Sensors and Actuators B: Chemical*, 255, pp. 3409–3416. Available at: https://doi.org/10.1016/j.snb.2017.09.170.
- Latu-Romain, L. *et al.* (2008) 'Growth parameters and shape specific synthesis of silicon nanowires by the VLS method', *Journal of Nanoparticle Research*, 10(8), pp. 1287–1291. Available at: https://doi.org/10.1007/s11051-007-9350-3.
- Law, M., Goldberger, J. and Yang, P. (2004) 'SEMICONDUCTOR NANOWIRES AND NANOTUBES', *Annual Review of Materials Research*, 34(1), pp. 83–122. Available at: https://doi.org/10.1146/annurev.matsci.34.040203.112300.
- Lawrence C. Gunn, Iiit.J.P.J.R.C. (2002) 'Waveguide photodetector with integrated electronics'. US. Available at: https://patents.google.com/patent/US7616904B1/en.
- Lazareva, E. and Tuchin, V. (2018) 'Blood refractive index modelling in the visible and near infrared spectral regions', *Journal of Biomedical Photonics & Engineering*, 4(1), p. 010503. Available at: https://doi.org/10.18287/JBPE18.04.010503.

- Leonardis, V.M.N.P.B.T.F. De (2012) 'Group IV Materials for High Performance Methane Sensing in Novel Slot Optical Waveguides at 2.883 µm and 3.39 µm', *Sensors & Transducers Journal*, 14(1), pp. 212–230.
- Li, H.H. (1976) 'Refractive index of alkali halides and its wavelength and temperature derivatives', *Journal of PhySical and Chemical Reference Data*, 5(2), pp. 329–528. Available at: https://doi.org/10.1063/1.555536.
- Li, K. *et al.* (2017) 'Integrated refractive index sensor using silicon slot waveguides', *Applied Optics*, 56(11), p. 3096. Available at: https://doi.org/10.1364/AO.56.003096.
- Li, Y. and Tong, L. (2008) 'Mach-Zehnder interferometers assembled with optical microfibers or nanofibers', *Optics Letters*, 33(4), p. 303. Available at: https://doi.org/10.1364/OL.33.000303.
- Liu, Q. *et al.* (2011) 'Dispersion engineering of a silicon-nanocrystal-based slot waveguide for broadband wavelength conversion', *Applied Optics*, 50(9), p. 1260. Available at: https://doi.org/10.1364/AO.50.001260.
- Liu, T. *et al.* (2013) 'Advances of optical fiber sensors for coal mine safety monitoring applications', in 2013 *International Conference on Microwave and Photonics (ICMAP)*. IEEE, pp. 1–5. Available at: https://doi.org/10.1109/ICMAP.2013.6733455.
- Loebner, E.E. (1976) 'Subhistories of the light emitting diode', *IEEE Transactions on Electron Devices*, 23(7), pp. 675–699. Available at: https://doi.org/10.1109/T-ED.1976.18472.
- Luo, Y. *et al.* (2014) 'Optofluidic glucose detection by capillary-based ring resonators', *Optics & Laser Technology*, 56, pp. 12–14. Available at: https://doi.org/10.1016/j.optlastec.2013.07.007.
- Lusse, P. *et al.* (1994) 'Analysis of vectorial mode fields in optical waveguides by a new finite difference method', *Journal of Lightwave Technology*, 12(3), pp. 487–494. Available at: https://doi.org/10.1109/50.285331.
- Maenaka, K. *et al.* (1996) 'Analysis of a highly sensitive silicon gyroscope with cantilever beam as vibrating mass', *Sensors and Actuators A: PhySical*, 54(1–3), pp. 568–573. Available at: https://doi.org/10.1016/S0924-4247(97)80016-7.
- Maese-Novo, A. *et al.* (2013) 'Wavelength independent multimode interference coupler', *Optics Express*, 21(6), p. 7033. Available at: https://doi.org/10.1364/OE.21.007033.
- Maine, P. *et al.* (1988) 'Generation of ultrahigh peak power pulses by chirped pulse amplification', *IEEE Journal of Quantum Electronics*, 24(2), pp. 398–403. Available at: https://doi.org/10.1109/3.137.
- Malinsky, M.D. *et al.* (2001) 'Chain Length Dependence and Sensing Capabilities of the Localized Surface Plasmon Resonance of Silver

Nanoparticles Chemically Modified with Alkanethiol Self-Assembled Monolayers', *Journal of the American Chemical Society*, 123(7), pp. 1471–1482. Available at: https://doi.org/10.1021/ja003312a.

- Marazuela, M. and Moreno-Bondi, M. (2002) 'Fiber-optic biosensors – an overview', *Analytical and Bioanalytical Chemistry*, 372(5–6), pp. 664–682. Available at: https://doi.org/10.1007/s00216-002-1235-9.
- Marcatili, E.A.J. (1969) 'Dielectric Rectangular Waveguide and Directional Coupler for Integrated Optics', *Bell System Technical Journal*, 48(7), pp. 2071–2102. Available at: https://doi.org/10.1002/j.1538-7305.1969.tb01166.x.
- Marchetti, R. *et al.* (2017) 'Group-velocity dispersion in SOI-based channel waveguides with reduced-height', *Optics Express*, 25(9), p. 9761. Available at: https://doi.org/10.1364/OE.25.009761.
- Martínez, R. V, Martínez, J. and Garcia, R. (2010) 'Silicon nanowire circuits fabricated by AFM oxidation nanolithography', *Nanotechnology*, 21(24), p. 245301. Available at: https://doi.org/10.1088/0957-4484/21/24/245301.
- McNichols, R.J. and Cote, G.L. (2000) 'Optical glucose sensing in biological fluids: an overview', *Journal of Biomedical Optics*, 5(1), p. 5. Available at: https://doi.org/10.1117/1.429962.
- Mehrotra, P. (2016) 'Biosensors and their applications – A review', *Journal of Oral Biology and Craniofacial Research*, 6(2), pp. 153–159. Available at: https://doi.org/10.1016/j.jobcr.2015.12.002.
- Melloni, A. *et al.* (2009) 'The role of index contrast in dielectric optical waveguides', *International Journal of Materials and Product Technology*, 34(4), p. 421. Available at: https://doi.org/10.1504/IJMPT.2009.024998.
- Mialdun, A. and Shevtsova, V. (2017) 'Analysis of multi-wavelength measurements of diffusive properties via dispersion dependence of optical properties', *Applied Optics*, 56(3), p. 572. Available at: https://doi.org/10.1364/AO.56.000572.
- Mosser, V. *et al.* (1991) 'Piezoresistive pressure sensors based on polycrystalline silicon', *Sensors and Actuators A: PhySical*, 28(2), pp. 113–132. Available at: https://doi.org/10.1016/0924-4247(91)85020-O.
- Mukundan, H. *et al.* (2009) 'Planar optical waveguide-based biosensor for the quantitative detection of tumor markers', *Sensors and Actuators B: Chemical*, 138(2), pp. 453–460. Available at: https://doi.org/10.1016/j.snb.2009.01.073.
- Nedeljkovic, M. *et al.* (2015) 'Surface-Grating-Coupled Low-Loss Ge-on-Si Rib Waveguides and Multimode Interferometers', *IEEE Photonics Technology Letters*, 27(10), pp. 1040–1043. Available at: https://doi.org/10.1109/LPT.2015.2405611.

- Nemova, G. and Kashyap, R. (2007) 'Theoretical model of a planar integrated refractive index sensor based on surface plasmon-polariton excitation with a long period grating', *Journal of the Optical Society of America B*, 24(10), p. 2696. Available at: https://doi.org/10.1364/JOSAB.24.002696.
- Orlando, A.J.F. (1992) 'Periodic Circular Waveguides Containing Azimuthally Magnetized Ferrite Rod', in *22nd European Microwave Conference,* 1992. IEEE, pp. 539–544. Available at: https://doi.org/10.1109/EUMA.1992.335761.
- Pan, H. *et al.* (2005) 'Growth of Si nanowires by thermal evaporation', *Nanotechnology*, 16(4), pp. 417–421. Available at: https://doi.org/10.1088/0957-4484/16/4/014.
- Peng, K. *et al.* (2005) 'Aligned Single-Crystalline Si Nanowire Arrays for Photovoltaic Applications', *Small*, 1(11), pp. 1062–1067. Available at: https://doi.org/10.1002/smll.200500137.
- Perez-Galacho, D. *et al.* (2013) 'Integrated polarization beam splitter with relaxed fabrication tolerances', *Optics Express*, 21(12), p. 14146. Available at: https://doi.org/10.1364/OE.21.014146.
- Pollock, C.R. and Lipson, M. (2003a) *Integrated Photonics*, *Springer US*. Boston, MA: Springer US. Available at: https://doi.org/10.1007/978-1-4757-5522-0.
- Pollock, C.R. and Lipson, M. (2003b) 'Rectangular Dielectric Waveguides', in *Integrated Photonics*. Boston, MA: Springer US, pp. 99–123. Available at: https://doi.org/10.1007/978-1-4757-5522-0_5.
- Priye, V., Malviya, N. and Mickelson, A. (2018) 'Analytical predictions for nonlinear optical processes in silicon slot waveguides', *Journal of Computational Electronics*, 17(2), pp. 857–865. Available at: https://doi.org/10.1007/s10825-018-1150-8.
- Qiu, T. *et al.* (2006) 'Intergrowth mechanism of silicon nanowires and silver dendrites', *Journal of Electronic Materials*, 35(10), pp. 1879–1884. Available at: https://doi.org/10.1007/s11664-006-0171-4.
- Raghuwanshi, S.K., Kumar, M. and Athokpam, B.S. (2016) 'Analysis of Novel Class of Surface Plasmon Phenomena Having a Metamaterial Layer Between Two Different Metals for Sensor Application', *IEEE Sensors Journal*, 16(17), pp. 6617–6624. Available at: https://doi.org/10.1109/JSEN.2016.2586541.
- Rahman, B. and Davies, J. (1984) 'Finite-element solution of integrated optical waveguides', *Journal of Lightwave Technology*, 2(5), pp. 682–688. Available at: https://doi.org/10.1109/JLT.1984.1073669.
- Reed, G.T. *et al.* (2010) 'Silicon optical modulators', *Nature Photonics*, 4(8), pp. 518–526. Available at: https://doi.org/10.1038/nphoton.2010.179.

- Reed, G.T., Headley, W.R. and Png, C.E.J. (2005) 'Silicon photonics: the early years', in J.A. Kubby and G.E. Jabbour (eds), p. 1. Available at: https://doi.org/10.1117/12.596921.
- Reed, G.T. and Knights, A.P. (2004) S*ilicon Photonics*. Chichester, UK: John Wiley & Sons, Ltd. Available at: https://doi.org/10.1002/0470014180.
- Rickman, A. (2014) 'The commercialization of silicon photonics', *Nature Photonics*, 8(8), pp. 579–582. Available at: https://doi.org/10.1038/nphoton.2014.175.
- Riordan, M., Hoddeson, L. and Herring, C. (1999) 'The invention of the transistor', *Reviews of Modern Phy*Sics, 71(2), pp. S336–S345. Available at: https://doi.org/10.1103/RevModPhys.71.S336.
- Robinson, S. and Dhanlaksmi, N. (2017) 'Photonic crystal based biosensor for the detection of glucose concentration in urine', *Photonic Sensors*, 7(1), pp. 11–19. Available at: https://doi.org/10.1007/s13320-016-0347-3.
- Ross, I.M. (1998) 'The invention of the transistor', *Proceedings of the IEEE*, 86(1), pp. 7–28. Available at: https://doi.org/10.1109/5.658752.
- Routkevitch, D. *et al.* (1996) 'Nonlithographic nano-wire arrays: fabrication, physics, and device applications', *IEEE Transactions on Electron Devices*, 43(10), pp. 1646–1658. Available at: https://doi.org/10.1109/16.536810.
- Rukhlenko, I.D., Premaratne, M. and Agrawal, G.P. (2012) 'Effective mode area and its optimization in silicon-nanocrystal waveguides', *Optics Letters*, 37(12), p. 2295. Available at: https://doi.org/10.1364/OL.37.002295.
- Ryckeboer, E. *et al.* (2014) 'Glucose sensing by waveguide-based absorption spectroscopy on a silicon chip', *Biomedical Optics Express*, 5(5), p. 1636. Available at: https://doi.org/10.1364/BOE.5.001636.
- Schmidt, V. *et al.* (2009) 'Silicon Nanowires: A Review on Aspects of their Growth and their Electrical Properties', *Advanced Materials*, 21(25–26), pp. 2681–2702. Available at: https://doi.org/10.1002/adma.200803754.
- Serna, S. *et al.* (2016) 'Experimental GVD engineering in slow light slot photonic crystal waveguides', *Scientific Reports*, 6(1), p. 26956. Available at: https://doi.org/10.1038/srep26956.
- Sharping, J.E. *et al.* (2005) 'All-optical, wavelength and bandwidth preserving, pulse delay based on parametric wavelength conversion and dispersion', *Optics Express*, 13(20), p. 7872. Available at: https://doi.org/10.1364/OPEX.13.007872.
- Silva, A.F. da *et al.* (2001) 'Thermal and optical properties of porous silicon', *Materials Research*, 4(1), pp. 23–26. Available at: https://doi.org/10.1590/S1516-14392001000100006.
- Singh, R.R. *et al.* (2019) 'Vertical silicon nanowire-based optical waveguide for DNA hybridization biosensor', in M. Razeghi and E. Tournie (eds)

Quantum SenSing and Nano Electronics and Photonics XVI. SPIE, p. 92. Available at: https://doi.org/10.1117/12.2507927.

- Singh, R.R. (2020a) 'Dispersion tailoring of silicon nanowire optical rectangular waveguide (SNORW)', *SN Applied Sciences*, 2(3), p. 502. Available at: https://doi.org/10.1007/s42452-020-2309-z.
- Singh, R.R. (2020b) 'Structural optimization and parametric analysis of SOI optical slot waveguides', *Journal of Computational Electronics* [Preprint]. Available at: https://doi.org/10.1007/s10825-020-01473-x.
- Singh, R.R., Malviya, N. and Priye, V. (2016) 'Parametric Analysis of Silicon Nanowire Optical Rectangular Waveguide Sensor', *IEEE Photonics Technology Letters*, 28(24), pp. 2889–2892. Available at: https://doi.org/10.1109/LPT.2016.2624501.
- Singh, R.R. and Priye, V. (2018) 'Silicon Nanowire Optical Rectangular Waveguide Biosensor for DNA Hybridization', *IEEE Photonics Technology Letters*, 30(12), pp. 1123–1126. Available at: https://doi.org/10.1109/LPT.2018.2835152.
- Sohi, P.A. and Kahrizi, M. (2017) 'Formation Mechanism of Silicon Nanowires Using Chemical/Electrochemical Process', *IEEE Transactions on Nanotechnology*, 16(3), pp. 507–513. Available at: https://doi.org/10.1109/TNANO.2017.2694428.
- Soler Penades, J. *et al.* (2014) 'Suspended SOI waveguide with sub-wavelength grating cladding for mid-infrared', *Optics Letters*, 39(19), p. 5661. Available at: https://doi.org/10.1364/OL.39.005661.
- Soref, R. (2006) 'The Past, Present, and Future of Silicon Photonics', *IEEE Journal of Selected Topics in Quantum Electronics*, 12(6), pp. 1678–1687. Available at: https://doi.org/10.1109/JSTQE.2006.883151.
- Soref, R. and Bennett, B. (1987) 'Electrooptical effects in silicon', *IEEE Journal of Quantum Electronics*, 23(1), pp. 123–129. Available at: https://doi.org/10.1109/JQE.1987.1073206.
- Soref, R.A. and Bennett, B.R. (1987) 'Kramers-Kronig Analysis Of Electro-Optical Switching In Silicon', in M.A. Mentzer and S. Sriram (eds), p. 32. Available at: https://doi.org/10.1117/12.937193.
- Soref, R.A. and Lorenzo, J.P. (1985) 'Single-crystal silicon: a new material for 1.3 and 1.6 μm integrated-optical components', *Electronics Letters*, 21(21), p. 953. Available at: https://doi.org/10.1049/el:19850673.
- Srivastava, S.K. *et al.* (2010) 'Excellent antireflection properties of vertical silicon nanowire arrays', *Solar Energy Materials and Solar Cells*, 94(9), pp. 1506–1511. Available at: https://doi.org/10.1016/j.solmat.2010.02.033.
- Steckl, A.J. *et al.* (2011) 'DNA as an Optical Material', *Optics and Photonics News*, 22(7), p. 34. Available at: https://doi.org/10.1364/OPN.22.7.000034.

- Steglich, P. (2018) ‘Silicon-on-Insulator Slot Waveguides: Theory and Applications in Electro-Optics and Optical Sensing’, in *Emerging Waveguide Technology*. InTech. Available at: https://doi.org/10.5772/intechopen.75539.
- Summers, L. *et al.* (1999) ‘The Use of Microdialysis to Monitor Rapid Changes in Glucose Concentration’, *Hormone and Metabolic Research*, 31(07), pp. 424–428. Available at: https://doi.org/10.1055/s-2007-978767.
- Sun, R. *et al.* (2007) ‘Horizontal single and multiple slot waveguides: optical transmission at λ = 1550 nm’, *Optics Express*, 15(26), p. 17967. Available at: https://doi.org/10.1364/OE.15.017967.
- Tamada, J.A., Bohannon, N.J.V. and Potts, R.O. (1995) ‘Measurement of glucose in diabetic subjects using noninvasive transdermal extraction’, *Nature Medicine*, 1(11), pp. 1198–1201. Available at: https://doi.org/10.1038/nm1195-1198.
- Tasdemir, Z. *et al.* (2015) ‘Monolithic Fabrication of Silicon Nanowires Bridging Thick Silicon Structures’, *IEEE Nanotechnology Express*, 1, pp. 2–5. Available at: https://doi.org/10.1109/XNANO.2015.2469312.
- Tasic, A.Z. *et al.* (1992) ‘Use of mixing rules in predicting refractive indexes and specific refractivities for some binary liquid mixtures’, *Journal of Chemical & Engineering Data*, 37(3), pp. 310–313. Available at: https://doi.org/10.1021/je00007a009.
- Thomson, D. *et al.* (2016) ‘Roadmap on silicon photonics’, *Journal of Optics*, 18(7), p. 073003. Available at: https://doi.org/10.1088/2040-8978/18/7/073003.
- Tien, E.-K. *et al.* (2010) ‘Discrete parametric band conversion in silicon for mid-infrared applications’, *Optics Express*, 18(21), p. 21981. Available at: https://doi.org/10.1364/OE.18.021981.
- Tong, J. *et al.* (2020) ‘Effects of the ambient medium and structure parameter on the optical properties of tapered silicon nanowire’, *Optics Communications*, 454, p. 124515. Available at: https://doi.org/10.1016/j.optcom.2019.124515.
- Van, V. (2017) *Optical Microring Resonators*. Boca Raton: CRC Press.
- Venkatesan, R. *et al.* (2020) ‘Investigating antireflection properties of hybrid silicon nanostructures comprising rod-like nanopores and nano-textured surface’, *Materials Letters*, p. 128087. Available at: https://doi.org/10.1016/j.matlet.2020.128087.
- Viphavakit, C. *et al.* (2015) ‘Optimization of a horizontal slot waveguide biosensor to detect DNA hybridization’, *Applied Optics*, 54(15), p. 4881. Available at: https://doi.org/10.1364/AO.54.004881.
- De Vos, K. *et al.* (2007) ‘Silicon-on-Insulator microring resonator for sensitive and label-free biosensing’, *Optics Express*, 15(12), p. 7610. Available at: https://doi.org/10.1364/OE.15.007610.

- Wageeh, A., El-Sabban, S. and Khalil, D. (1995) 'Resonance Wavelength of Integrated Optical Ring Resonator With Small Radius of Curvature', *IEEE Photonics Technology Letters*, 26(7), pp. 641–644. Available at: https://doi.org/10.1109/LPT.2014.2301755.
- Wagner, R.S. and Ellis, W.C. (1964) 'VAPOR-LIQUID-SOLID MECHANISM OF SINGLE CRYSTAL GROWTH', *Applied PhySics Letters*, 4(5), pp. 89–90. Available at: https://doi.org/10.1063/1.1753975.
- Wagner, R.S. and Ooherty, C.J. (1968) 'Mechanism of Branching and Kinking during VLS Crystal Growth', *Journal of The Electrochemical Society*, 115(1), p. 93. Available at: https://doi.org/10.1149/1.2411032.
- Wang, J., Glesk, I. and Chen, L.R. (2014) 'Subwavelength grating filtering devices', *Optics Express*, 22(13), p. 15335. Available at: https://doi.org/10.1364/OE.22.015335.
- Wang, N. *et al.* (1998) 'SiO2-enhanced synthesis of Si nanowires by laser ablation', *Applied PhySics Letters*, 73(26), pp. 3902–3904. Available at: https://doi.org/10.1063/1.122930.
- Wang, N. *et al.* (1999) 'Si nanowires grown from silicon oxide', *Chemical PhySics Letters*, 299(2), pp. 237–242. Available at: https://doi.org/10.1016/S0009-2614(98)01228-7.
- Wang, Q., Farrell, G. and Freir, T. (2006) 'Effective index method for planar lightwave circuits containing directional couplers', *Optics Communications*, 259(1), pp. 133–136. Available at: https://doi.org/10.1016/j.optcom.2005.08.054.
- Weakliem, H.A. and Redfield, D. (1979) 'Temperature dependence of the optical properties of silicon', *Journal of Applied PhySics*, 50(3), pp. 1491–1493. Available at: https://doi.org/10.1063/1.326135.
- Westerveld, W.J. *et al.* (2012) 'Extension of Marcatili's Analytical Approach for Rectangular Silicon Optical Waveguides', *Journal of Lightwave Technology*, 30(14), pp. 2388–2401. Available at: https://doi.org/10.1109/JLT.2012.2199464.
- Westwater, J. (1997) 'Growth of silicon nanowires via gold/silane vapor–liquid–solid reaction', *Journal of Vacuum Science & Technology B: Microelectronics and Nanometer Structures*, 15(3), p. 554. Available at: https://doi.org/10.1116/1.589291.
- Xiong, Y. *et al.* (2014) 'Polarization splitter and rotator with subwavelength grating for enhanced fabrication tolerance', *Optics Letters*, 39(24), p. 6931. Available at: https://doi.org/10.1364/OL.39.006931.
- Xiong, Y. *et al.* (2015) 'High Extinction Ratio and Broadband Silicon TE-Pass Polarizer Using Subwavelength Grating Index Engineering', *IEEE Photonics Journal*, 7(5), pp. 1–7. Available at: https://doi.org/10.1109/JPHOT.2015.2483204.

- Xu, Q. *et al.* (2004) 'Experimental demonstration of guiding and confining light in nanometer-size low-refractive-index material', *Optics Letters*, 29(14), p. 1626. Available at: https://doi.org/10.1364/OL.29.001626.
- Yahya, M. and Saghir, M.Z. (2016) 'Empirical modelling to predict the refractive index of human blood', *PhySics in Medicine and Biology*, 61(4), pp. 1405–1415. Available at: https://doi.org/10.1088/0031-9155/61/4/1405.
- Yan, R., Gargas, D. and Yang, P. (2009) 'Nanowire photonics', *Nature Photonics*, 3(10), pp. 569–576. Available at: https://doi.org/10.1038/nphoton.2009.184.
- Yang, X.C. *et al.* (2016) 'A photonic crystal fiber glucose sensor filled with silver nanowires', *Optics Communications*, 359, pp. 279–284. Available at: https://doi.org/10.1016/j.optcom.2015.09.102.
- Yang, Y.-H. *et al.* (2004) 'Catalytic Growth of Silicon Nanowires Assisted by Laser Ablation', *The Journal of PhySical Chemistry B*, 108(3), pp. 846–852. Available at: https://doi.org/10.1021/jp030663d.
- Yang, Z. *et al.* (2001) 'Simulation of optical waveguides with FDTD method', in J. Liu and Z. Wang (eds) *Proceedings of SPIE - The International Society for Optical Engineering*, p. 42. Available at: https://doi.org/10.1117/12.444528.
- Yeh, Y.-L. (2008) 'Real-time measurement of glucose concentration and average refractive index using a laser interferometer', *Optics and Lasers in Engineering*, 46(9), pp. 666–670. Available at: https://doi.org/10.1016/j.optlaseng.2008.04.008.
- Yoon, J.-S. *et al.* (2017) 'Optical Characteristics of Silicon-Based Asymmetric Vertical Nanowire Photodetectors', *IEEE Transactions on Electron Devices*, 64(5), pp. 2261–2266. Available at: https://doi.org/10.1109/TED.2017.2682878.
- Yuan, S.-N. and Fang, Y.-T. (2017) 'Competition and evolution of dielectric waveguide mode and plasmonic waveguide mode', *The European PhySical Journal Applied PhySics*, 80(2), p. 20501. Available at: https://doi.org/10.1051/epjap/2017170241.
- Yue, Y. *et al.* (2010) 'Highly efficient nonlinearity reduction in silicon-on-insulator waveguides using vertical slots', *Optics Express*, 18(21), p. 22061. Available at: https://doi.org/10.1364/OE.18.022061.
- Yun, H. *et al.* (2016) 'Broadband 2 × 2 adiabatic 3 dB coupler using silicon-on-insulator sub-wavelength grating waveguides', *Optics Letters*, 41(13), p. 3041. Available at: https://doi.org/10.1364/OL.41.003041.
- Zengzhi, H. *et al.* (2015) 'High Confinement Factor Ridge Slot Waveguide for Optical Sensing', *IEEE Photonics Technology Letters*, 27(22), pp. 2395–2398. Available at: https://doi.org/10.1109/LPT.2015.2466595.

- Zhang, L. *et al.* (2009) 'Highly dispersive slot waveguides', *Optics Express*, 17(9), p. 7095. Available at: https://doi.org/10.1364/OE.17.007095.
- Zhang, L., Yue, Y., Xiao-Li, Y., *et al.* (2010) 'Flat and low dispersion in highly nonlinear slot waveguides', *Optics Express*, 18(12), p. 13187. Available at: https://doi.org/10.1364/OE.18.013187.
- Zhang, L., Yue, Y., Beausoleil, R.G., *et al.* (2010) 'Flattened dispersion in silicon slot waveguides', *Optics Express*, 18(19), p. 20529. Available at: https://doi.org/10.1364/OE.18.020529.
- Zhang, L. *et al.* (2012) 'Silicon waveguide with four zero-dispersion wavelengths and its application in on-chip octave-spanning supercontinuum generation', *Optics Express*, 20(2), p. 1685. Available at: https://doi.org/10.1364/OE.20.001685.
- Zhang, Y. *et al.* (2015) 'Supercontinuum generation in strip/slot hybrid waveguide with flat and low dispersion', *Applied Optics*, 54(15), p. 4850. Available at: https://doi.org/10.1364/AO.54.004850.
- Zhang, Y.F. *et al.* (2000) 'Bulk-quantity Si nanowires synthesized by SiO sublimation', *Journal of Crystal Growth*, 212(1–2), pp. 115–118. Available at: https://doi.org/10.1016/S0022-0248(00)00238-4.
- Zhao, C.Z. *et al.* (1995) 'Silicon on insulator Mach–Zehnder waveguide interferometers operating at 1.3 μm', *Applied Phy*Sics *Letters*, 67(17), pp. 2448–2449. Available at: https://doi.org/10.1063/1.114603.
- Zheng, Z., Iqbal, M. and Liu, J. (2008) 'Dispersion characteristics of SOI-based slot optical waveguides', *Optics Communications*, 281(20), pp. 5151–5155. Available at: https://doi.org/10.1016/j.optcom.2008.07.003.
- Zhou, F. *et al.* (2020) 'Temperature Insensitive Fiber Optical Refractive Index Probe with Large Dynamic Range at 1,550 nm', *Sensors and Actuators A: Phy*Sical, p. 112102. Available at: https://doi.org/10.1016/j.sna.2020.112102.
- Zhu, M. *et al.* (2012) 'Ultrabroadband flat dispersion tailoring of dual-slot silicon waveguides', *Optics Express*, 20(14), p. 15899. Available at: https://doi.org/10.1364/OE.20.015899.

❑❑❑

Index

For Product Safety Concerns and Information please contact our EU representative GPSR@taylorandfrancis.com
Taylor & Francis Verlag GmbH, Kaufingerstraße 24, 80331 München, Germany

www.ingramcontent.com/pod-product-compliance
Ingram Content Group UK Ltd.
Pitfield, Milton Keynes, MK11 3LW, UK
UKHW022313250926
475702UK00046B/763